EARTHFORCE!

AN EARTH WARRIOR'S GUIDE TO STRATEGY

2nd Edition

CAPTAIN PAUL WATSON

Earthforce!

Second Edition Copyright © 2012 Paul F. Watson

All rights reserved.

Cover illustration by
Tiphaine Blot & Jérémy Charrance ©Altering Studios 2012
Cover graphics by Céline Tremblay
Edited by Tiffany Humphrey

ISBN-13: 978-1-61419-016-5

I would like to dedicate this book to Edward Abbey, Cleveland Amory, Brigitte Bardot, Bob Barker, Robert Hunter, Farley Mowat and Steve Irwin whose names have graced the bows of Sea Shepherd ships over the years.

CONTENTS

Acknowledgments	vi
Foreword by Richard Dean Anderson	vii
Introduction	1
Principal Contributors	7
Foundation	16
Earthforce - The Spiritual Foundation of an Earth Warrior	27
The Continuum	31
Chapter 1 Preparations	39
Chapter 2 Deception	53
Chapter 3 Confrontation	60
Chapter 4 The Art of Fighting Without Fighting	65
Chapter 5 Tactics	70
Chapter 6 Earthforce	73
Chapter 7 Stealth and Unpredictability	77
Chapter 8 Maneuvering	81
Chapter 9 Maintaining the Center	84
Chapter 10 Field Confrontation	86
Chapter 11 Terrain And Situations	89
Chapter 12 Attacking With Fire	93
Chapter 13 The Use of Intelligence Agents	96
The 36 Strategies	101
Spaceship Earth	107
Afterwords	111

ACKNOWLEDGMENTS

I'd like to thank Tiphaine Blot and Jérémy Charrance at Altering Studios for creating the cover artwork and Celine Tremblay for the cover graphics. I'd also like to thank Tiffany Humphrey, my assistant, for shepherding this project to completion.

Foreword

I'd like to throw a huge spotlight in the direction of a fellow humanoid whose problems, distractions, and walls of intolerance appear insurmountable. There are people, hell, virtually entire industries and governments who openly despise Paul Watson. We must all know by now, the openly defiant Japanese Whaling industry, is very near poised to bomb any one of three ships out of the water or shoot down the surveillance helicopter that regularly buzzes what the whalers consider, "their space".

That there is any need for Captain Paul Watson and his devoted crew to be in Antarctica AT ALL is an absolutely maddening question to all of us who want and need to support Sea Shepherd and Captain Watson; to those of us who actually have an ounce of respect, compassion, understanding, and willingness to acknowledge the universe has changed, and will forever be doing so; to those of us willing to make any and all necessary changes in culture, daily living, and personal desires that will contribute to Earths' ultimate rescue.

It might seem ironic to consider meeting a man of the sea whilst mucking about at altitude, some 12,000 feet above the sea's level. My first encounter with Paul was in an environment far more familiar to me than the endlessness that IS the sea. We were both invited to a fundraising event benefiting the Sea Shepherd Conservation Society (SSCS), north of Anchorage, Alaska. I had become a semi-regular with the Marjoe Gortner group, who, every year, would "throw" an event for any cause deemed worthy of Marjoes' generous intentions. All manner of celebrity-type people would be gifted a miniature vacation at some distant resort, all expenses paid and lots and lots of sponsor product (SWAG, I think it is called).

The event featured a "final contest" 'twixt warring teams, each Captained by a biggish celebrity and at the ski events, usually a US Ski Team alum, or Olympian still capable of smoking anybody down a hill and through the gates of a very conservatively set-up race course. It was always a barrel of fun. Unless, of course, some celeb-type mogul took it all way too seriously, who really HAD to win with the snazziest team creds and a guarantee of taking home yet another present.

It was at this particular Alaskan event that I met Captain Paul Watson. The irony was not lost on me; frolicking through some of the most glorious of mountain environments in Alaska, all of it to make some money for the Sea Shepherd Conservation Society, and Captain Paul Watson. The Gortner group was honoring Sea Shepherd and presenting most (if not some) of the funds garnered during the weekend, to Captain Paul, head shepherd of the organization, and as far as Marjoe Gortner was concerned, a rather mysterious and vague individual.

I think most peoples' first impression of Paul isn't necessarily the warmest; although, I must admit I have always had a burning desire to hug the Dickens out of him, he's just that damn cute. Cutting to the point here, the bottom line about MY first impression of Paul was…..well, here is a rather large fellow! Solid man, for sure, and donning a beautiful mop of silvery-white hair. As was customary, the beneficiary of the event's monies, (donations corporate and individual, silent auction, regular, not-so-silent auction, and sponsor dollars from consistently generous "regulars", like John Paul DeJoria, Ron Rice, wine merchants, equipment venders, etc.) was expected to "say a few words," of informative inspiration. In Paul's case, he pretty much had to tell us who, what, where and why Sea Shepherd… was!

This was no easy task in those days. My recollection has it that, Greenpeace was the go-to group for all things oceanic, pretty much owning the "protest" crown. OH, how irony can get lost!!! As Paul would later impart: he co-founded Greenpeace, and it was therefore OK for him to say that Greenpeace* had become too full of itself, and it was becoming more and more difficult to get anything accomplished… especially on behalf of the marine animals who, as had become disturbingly apparent, needed DIRECT INTERVENTION, someone or something to stand in the way of the shower of harpoons that seemed to continue despite GP's spray-painted signs of protest….In a nut: Paul moved on to Greener pastures with his new baby, Sea Shepherd.

My interpretation got retained as: Greenpeace now existed to perpetuate itself. Don't get me wrong, "Greenpeace" is a terrific name for a conservation company, and in their day I thought they were very cool... But now, I keep waiting for them to DO something! As I draw myself back from the tangential thought form that is my anchor to bear, I am remembering the exact moment I became a devotee to Paul and the SSCS. It had become so starkly apparent that I was listening to a highly intelligent humanoid that particular evening in Alaska. His knowledge of all things under the sea, is so frighteningly vast, and is supported by explosive compassion and all the rites (rights) given to a man of endless experience, who has borne witness to the most horrific, blood-stained acts of man.

That evening, Paul told his audience a story; it is a story that I have now heard many, many times. It imparts the details of his encounter with a mother sperm whale, who was positioning herself in the water between a whaler's harpoon boat and her calf. Paul, in his own small boat, had been attempting to position himself between the whaler's deadly harpoon and the mother whale. When you get the chance, ask Paul to tell the story to you; I can't do it justice. And, besides............ you kind of had to be there. What I know for sure is, I listened to Paul Watson speak about his passion, with passion, and the experience has positively altered the course of my life.

I will forever be indebted (,Captain).

-Richard Dean Anderson

INTRODUCTION

EARTHFORCE!
An Earth Warrior's Guide to Strategy

"Yep, son, we have met the enemy and he is us."

- Walt Kelly
(Spoken by Pogo, the fictional opossum in the Okefenokee swamp)

I first wrote *Earthforce! A Guide to Strategy for the Earth Warrior* in 1993, the year after attending the 1992 United Nations Conference on the Environment and Development in Rio de Janeiro, Brazil. It has now been nineteen years since the first edition was published and much has changed. It is time for this little booklet to re-emerge once again. None of the promises made by all the nations that attended the Brazil conference have materialized but nonetheless they will meet again in Rio in 2012 to pretend yet once again to resolve the threats to the environment that loom larger and larger with each passing year.

Many of my predictions from 1993, such as the collapse of worldwide fisheries, the collapse of coral reef eco-systems and climate change have unfortunately come to pass, and finally, there is now a growing consciousness that the Earth is experiencing some very serious and significant ecological threats. To address these threats, there is a need for new ideas, new strategies, new tactics, and a renewed worldview.

There is a growing awareness of the threats to our planet and us, but the selfishness and greed of humanity fuelled by denial and special interests has seen very little progress on the part of governments, corporations and large NGO's to affect any positive change.

All the international conferences in Cairo, Copenhagen, Kyoto etc., on issues of climate change, population and sustainability have failed to go any further than the signing of international protocols and agreements between nations by leaders who want to be seen as addressing the problems but who lack the political courage and ecological education to implement the requirements outlined in the agreements.

Where change IS happening is with the passion, imagination and courage of individuals, community groups and small non-governmental organizations.

I have had the opportunity to observe the growth of the environmental movement during the last two decades. I have seen many mistakes and also many exciting and innovative ideas and initiatives. The second edition of this book will allow me the opportunity to expand these ideas on strategy in a more detailed and up to date manner.

The environmental and conservation movement has grown stronger each year since I co-founded the Greenpeace Foundation in 1972. That was also the year we placed twelve billboards in Vancouver with the word ECOLOGY in large letters. And underneath we had a short sentence that read – Look it up! You're involved.

There are few people today who do not at least know what that word means and a growing number of people, at least in the developed world, are well aware that the threat to the environment, and thus to their own future, is very real.

Environmentalism is the fastest growing movement in the world today and it will soon become the single most important social movement in the history of human civilization. This will not be by choice, but by necessity.

As I wrote in 1993, becoming an environmentalist is in reality a question of self-preservation.

Ecological systems are simply life support systems for Spaceship Earth. If these systems fail, then the consequences are fatal and permanent, for all of humanity, and most all other species.

Defending the natural world will become the challenge for a new type of Warrior. For many years I have envisioned the development of a philosophy and a discipline to be defined as the Way of the Earth Warrior.

If this planet is to be saved from ourselves we need to construct a Warrior Society of courageous, passionate, imaginative and resourceful men and women, willing to take the risks required to make a difference in the struggle to serve and protect our common mother – the Earth.

This book was originally published to be a guide to strategy for people who desire to make a difference, who seriously wish to be a part of the solution and not part of the problem. Nothing has changed, it remains a book for people who are ready to stand up for what they stand upon – the Earth beneath their feet. The intent of the book is the same, only the strategies and philosophy have been updated.

It is my belief that what is needed is a philosophy that is biocentric in its approach as opposed to anthropocentric and that provides guidance for people wishing to work and fight for a world that sees the species *homo sapiens* living in harmony with the natural world in accordance with the laws of ecology.

The modern environmentalist philosophy had its roots in the writings and actions of Henry David Thoreau and John Muir in the 19th Century. The German naturalist Ernst Haeckel invented the word "ökologie" in 1866 to describe his idea of a planetary "household" and an inter-connected family of species.

The movement was invigorated in the mid 20th Century by the writings and actions of Aldo Leopold, Rachel Carson, Edward Abbey, Farley Mowat, David Brower, Arne Naess, David Foreman, Robert Hunter, Dr. David Suzuki, Paul and Anne Ehrlich and others.

In the 1960's, there was a realization that ecology was more than just a science.

In the journal Bioscience, in 1964, Paul Sears called ecology "the subversive subject." He could see that the true implications of the understanding of ecology could radically change human society, health, politics, economics- in fact everything. Ecology represented the key to overthrowing the dominant paradigm. In other words, an ecological movement represented the most dominant subversive threat to the status quo ever seen.

It has been my lifelong objective to use ecology to overthrow the tyranny that anthropocentrism has held over the natural world. To me ecology is a science, but it is also a set of laws and it is the key to saving this wonderful jewel of a planet from ourselves and the ecological insanity of humanity. It is the only science that can form the basis for a global political movement.

This book will seek to describe just how the science of ecology sets forth the basic laws of ecology and can be utilized in strategy to achieve solid logistical objectives.

The roots of modern direct action environmentalism began with myself and the other co-founders of the Greenpeace Foundation in

1969 when we formed the Don't Make a Wave Committee and organized the voyage in 1971 to oppose nuclear testing on the Aleutian island of Amchitka. In 1972, we changed the name of the Don't Make a Wave Committee to establish the Greenpeace Foundation. I was the youngest founding member and director with the lifetime membership number of 007 making me the 8th founding member. Robert Hunter, who became my lifelong friend and often partner in direct-action conservation until his death in 2005, had the number of 000.

This movement became more aggressive when I founded the Sea Shepherd Conservation Society in 1977, and Dave Foreman, Howie Wolke, and Mike Roselle founded Earthfirst! in 1980.

Since the beginning of the Eighties, environmentalists and conservationists have been jailed, beaten, and murdered. This has tempered the environmental movement with the blood and sacrifice of martyrs.

The roster of murdered martyrs include Joy Adamson (1980), Dian Fossey (1985), Fernando Pereira (1985), Chico Mendes (1988), and George Adamson (1989) among others.

Environmentalism is the most universal, noble, just, and moral of causes, and most importantly, represents the interests of every living thing upon the Earth. The enemy may very well be ourselves, but our allies are also ourselves. We choose to side with the Earth for the long term. The alternative is to side with the anthropocentric priorities of humanity in the short term.

The environmental movement is also a movement of complex diversity and just as any ecosystem is strengthened by diversity so also are social movements made strong by diversity. Diverse strategies and tactics all can find a constructive niche within the complexity of the overall spectrum of the ecology movement. Politically there is no right or left because the consequences affect all of humanity and the solutions benefit all of humanity in the most basic way.

This book is meant to be a guide for strategy. In any new and growing movement mistakes are made, sometimes with very serious consequences. The movement to defend planetary ecosystems is still without much direction, solid leadership, tested tactics, and is very weak in strategy.

The first step in formulating a comprehensive strategy is to ask the following questions.
Who are we?
What do we want?
Who are our enemies?
Who are our allies?
Who do we fight?
Who do we defend ourselves from?
Who do we ignore?
Where are we going?
How are we going to get there?
What do we do when we realize our objectives?
There are many questions and few answers.

It is my intention with this book to present guidelines which may help in answering some of these questions.

This is a book about strategy and how to apply strategy in the environmental and conservation movements. It is also applicable to the animal rights and human rights movements.

Strategy is by nature a martial discipline. In preparing this book I have drawn on four sources:

The first and most important book is a little tome written twenty-five centuries ago in China by a general named Sun Tzu. The book, entitled *The Art of War*, has been considered by many strategists throughout the centuries as the best work on martial strategy ever written. It has had a significant impact on political and military events throughout the history of civilization. I have applied Sun Tzu's basic principles to the development of this manifesto on ecological strategy.

The second source is another small book written in 1648 in Japan by a warrior/philosopher named Miyamoto Musashi. The work is entitled *The Book of Five Rings*. Musashi adds to the importance of Sun Tzu's work by providing guidance directly to individuals within a movement.

The third source is the written and presented works of Marshall McLuhan, the Canadian philosopher of media. In our present culture, it is the mass media which determine the boundaries of what we perceive as reality. It is within the media-defined culture that the environmental movement can do battle most effectively. An understanding of these media, which dominate global culture, is imperative for the implementation of an effective strategy.

The fourth source, humble in comparison to the first three, is my own experiences as an environmental activist and field campaign leader for the past forty years.

I have attempted to apply in the field, the lessons provided by the other three. I have had failures only where I neglected to adhere to the strategy guided by the principles of these three great teachers.

Since the original publication in 1993, I have led numerous sea-going campaigns against Norwegian, Faeroese, and Icelandic whaling operations in the North Atlantic and eight expeditions to the Southern Ocean to confront illegal Japanese whaling operations. In addition we have taken on the dolphin killers of Taiji, Japan, the seal killers off the Eastern Canadian coast and the coast of Namibia. We have confronted the Makah tribal whale hunt off the coast of Washington. We have set up a permanent mission in the Galapagos to work with the police and the rangers to control poaching in the Galapagos National Park Marine Reserve. These are operations that have involved numerous ships and hundreds of crewmembers in remote and hostile waters on a global scale.

My laboratory for the study of strategy has been these missions going back to 1969, more than forty years of confrontations in the field, in the courts, in the media and within the court of public opinion. During this entire time, I have been jailed but never convicted of a felony, and most importantly I have successfully been able to accomplish my goals without causing a single injury to any of the enemies I have opposed.

I hope that my experience will serve to assist others in the development and deployment of effective strategies for intelligent involvement and action.

It is my intention that this book will serve to motivate others to expand on strategic ideas for making the environmental movement stronger, more active, more directed, and much more effective. It is also my hope that this book will continue to inspire individual initiative and to help individuals to guide their passions with a clear and practical strategy.

PRINCIPAL CONTRIBUTORS

Sun Tzu
A wood-carved portrait of Sun Tzu from a thousand-year-old book. Artist unknown.

Miyamoto Musashi
Painting by the Japanese artist Kuniyoshi, c. 1848.

Marshall McLuhan
Photograph by Corinne McLuhan.

Paul Watson
Portrait by Sam Morse-Brown, official artist for the British War Museum, onboard the Sea Shepherd II *in Bermuda, 1979.*

PRINCIPAL CONTRIBUTORS

SUN TZU

In China, *The Art of War* has been known for centuries as The Thirteen Chapters. The author credited with writing this work is known as Sun Tzu or Sun Wu. Very little is known about him and some scholars have questioned if he was in fact one man or a fictional representation of more than one general.

An analysis of the weaponry and political situations described in the book does narrow the period of the authorship to a general time frame. Since the use of crossbows is mentioned, the book was probably written after 400 B.C., which is approximately when the crossbow was introduced into China. The fact that the book mentions large armies and the use of infantry but says not a word about cavalry suggests that the book was written before 320 B.C., which is the time King Wu Ling of Chao State introduced mounted troops.

The question of ancient literary authorship is always open to dispute by academicians. The controversy over the facts about the existence of Sun Tzu has been raging and smoldering for centuries and, likely will continue for centuries more.

What is certain is that *The Art of War* was written or edited by a unique and imaginative individual who was well versed in the practical experience of field warfare and conflict.

The text suggests an officer of high rank with exceptional skill in marshalling large numbers of forces in confrontation with formidable opposition.

The man, whatever his real name, was a gifted leader and one of the greatest of generals.

In 1782, *The Art of War* was translated into French by Father Amiot, a Jesuit. The Jesuits introduced the book to Napoleon Bonaparte. There are many historians who believe that The Thirteen Chapters was the secret to the success of the famed French Emperor. We know that Napoleon relied heavily on mobility, a technique stressed by Sun Tzu.

A careful study of Bonaparte's campaigns reveals that Sun Tzu's tactics were brilliantly employed by Le Grand Emperor. It is probable that in the end, he ignored The Thirteen Chapters in favour of a belief in his own invincibility.

The Vietnam War is a textbook example of what not to do according to the teachings of *The Art of War*. That war was protracted and the supply lines were too long. The United States failed to understand the strategy of the Viet Cong (VC) and the North Vietnamese Army (NVA). General Giap of the NVA read Sun Tzu thoroughly. The West Point officers who opposed him did not include The Thirteen Chapters in their curriculum. The outcome of the war in Indo-China was predicted in advance by students of Sun Tzu.

The United States is a country that traditionally wins with weight, not strategy, like a sumo wrestler sitting on a dwarf. This approach has worked against countries like Panama and Grenada but it is a limited strategy. U.S. foreign policy makes little attempt to employ stratagems for peace. The nation reacts to outside stimuli not unlike an amoeba reacts to light. The strategy of the United States government has traditionally failed to recognize the ultimate objective of Sun Tzu's philosophy of strategy. Thus, the U.S. has become bogged down in protracted hugely expensive wars with high civilian body counts in Vietnam, Iraq and Afghanistan, struggles that they cannot win. In fact, the U.S. could not even triumph over Somalia.

American style warfare has been overkill. In fact, a contributing factor to the rise of Nazism was the fact that General John Pershing threw so much weight behind the French and British forces in the First World War that the balance of power was so offset that the French were able to humiliate the Germans. If not for the Americans, the Treaty of Versailles would have been an armistice of near equality and the conditions for the rise of Nazism would not have occurred. This was a failure of foresight on the part of President Woodrow Wilson. Victory through humiliation has its consequences. Strategy must be far-seeing and not just immediate.

Examples of effective U.S. military interventions can be found in the activities of U.S. Navy Seals, Delta Force or the U.S. Rangers; Elite, highly mobile, strongly supported fighting groups utilized for surgical strikes.

It was a special Navy Seals force that found and took out Osama Bin Laden for example, not the shock and awe tactics of President

George Bush and an invading army into a nation completely unconnected the Bin Laden.

Large-scale grossly expensive invasion forces have not accomplished much of value except to brutalize the nation invaded, and to vastly increase the debt of the invading nation.

THE OBJECTIVE OF STRATEGY IS PEACE.

The United States did not and will not defeat communism. Communism defeated itself, just as that other anti-nature economic system called capitalism will eventually do to itself. Both economic systems have little regard for long term objectives and consequences.

For the environmental movement to triumph over the enemies of the Earth, it will be imperative to develop and implement an effective overall approach, which contains numerous supporting strategies and a long term vision.

What cannot be achieved with wealth needs to be achieved by stealth. What cannot be achieved by stealth needs to be achieved by profile and media, by the strength and loyalty within the movement, by courage, passion and leadership, and finally by the moral superiority of the cause.

Sun Tzu was a man of war whose objective was peace. The environmental movement is essentially a struggle to achieve ecological peace and harmony.

Sun Tzu wrote, "the true object of war is peace."

For the ecological warrior, the objective is ecological peace.

The path to that peace lies in strategy.

MIYAMOTO MUSASHI

The British have their most legendary historical figure in the personality of Robin Hood: independent, antiauthoritarian, courageous, ingenious, possessor of a strong moral code, and the personification of a self-disciplined independent warrior.

The Japanese have all of these traits and more in the person of Miyamoto Musashi, who was a man of history, shrouded in the mantle of glorious legends, all of which have a foundation in actual fact.

Miyamoto Musashi was born in about the year 1584 and is believed to have died in 1645. He was a master swordsman, probably the greatest master of kendo in all of Japanese martial history, and he developed the method of fighting with two swords. He was a rigid adherent to strict self-discipline as the key to martial arts.

As a youth, Musashi fought on the losing side in the famous battle of Sekigahara. After that he lived the life of a ronin or masterless samurai, which gave him both the freedom and the time to devote himself to the mastery of the art of the sword. He clashed with and defeated the masters of the Yoshioka School of swordsmanship in Kyoto and the monks of the Hozoin in Nara.

His most famous duel was his triumph over the sword master Sasaki Kojiro. He fought on the losing side at the siege of Osaka Castle in 1614 and participated in the 1637-38 annihilation of Christian forces at Shimabara on the western island of Kyushu, an event which resulted in the banishment of Christianity as a religion from Japan for the next two centuries.

In 1640, he became a retainer of the Hosokawa lords of Kumamoto. He was a skilled painter, poet, and sculptor as well as a swordsman.

Miyamoto Musashi left us with his legacy in the book entitled *Gorin No Sho* which translates as *The Book of Five Rings*.

Unlike *The Art of* War, which is a book of general strategy, the *Gorin No Sho is a book* of strategy aimed at the individual.

The author was a man who had transformed himself from a purely instinctive fighter into a warrior who pursued the goals of rigid self-discipline. He developed complete mastery over himself which was accompanied by a sense of oneness with nature. Through Musashi, martial skills, spiritual self-discipline, and aesthetic sensitivity were merged into a single harmonious whole.

Musashi offers guidance for the individual warrior within the Earthforce.

MARSHALL McLUHAN

Herbert Marshall McLuhan was born July 21, 1911 in Edmonton, Alberta.

Professor McLuhan taught at the University of Toronto from 1946 until his death on December 31, 1980, as a professor of

Literature and director of the University's Centre for Culture and Technology.

Marshall McLuhan's perceptions of the media have been documented in his books beginning with *The Mechanical Bride: Folklore of Industrial Man,* which was published in 1951. His second book, published in 1964, was *The Gutenberg Galaxy: The Making of Typographic Man,* followed by *Understanding Media: The Extensions Of Man* in 1964 and *The Medium is the Massage: An Inventory of Effects* in 1967.

McLuhanism is the name given to the media philosophy developed by Marshal McLuhan. The utilization of McLuhan's philosophy in dealing with the media is called a McLuhanesque approach or strategy.

Briefly and simply, McLuhan's understanding of the media is thus: the medium is the message. Each medium encourages some approaches to communication and discourages others, regardless of the subject. The media defines culture, and thus determines reality as it is perceived by human populations and cultures.

A McLuhanesque approach to the media is to understand what the media wants and then deliver your message to the selected media. If your media is television then you must exploit images and talk in sound bites. If your media is newspapers, then you give more attention to fact as well as attempt to influence the headline.

McLuhan describes various media as "hot" or "cool". Television is a "cool" medium because of its relatively "low definition" which engages its audience in an active way. Print is a "hot" medium, in that its "high definition" encourages detachment. For instance, a war being a hot issue is perceived emotionally on a "cool" media like television but can be read with detachment from a "hot" medium like a newspaper.

The media is an extension of the body. Newspapers and books are extensions of the eye. Electronic media, however, are an extension of the entire nervous system. Newspapers inform. Television and film entertains and enthralls.

To be successful, the modern Earth Warrior must be in control of the forces that define the dominant culture. Because the media defines culture, an understanding of media philosophy is essential.

CAPTAIN PAUL FRANKLIN WATSON

I was born on December 2, 1950 in the hereditary lands of the Huron on the north shore of Lake Ontario. I was raised in the east in the lands of the Alquonquin Micmac on the shores of the Passamaquoddy Bay. For the last forty years, I have made my home in the traditional lands of the West Coast Salish peoples. I remain linked in spirit to those I fought with at Wounded Knee in the lands of the Oglala Lakota. And I remain linked in spirit to the Mohawk people of Kahnawake for presenting me with a traditional shirt and the right to fly the flag of the Five Nations of the Iroquois from our ships.

During the month of March, 1973, I served as a medic with the American Indian Movement and participated in the defence of Wounded Knee. That month gave me my first and only military combat experience under fire. We held out against the superior forces of the United States government for more than two months.

I was fortunate in having the opportunity to run off to sea at an early age. This gave me first hand experience with numerous cultures including the witnessing of the armed revolution by FRELIMO in Mozambique and the anti-apartheid movement in South Africa. I travelled extensively throughout East and Southern Africa, the Middle East, Europe, and Asia. My sea experience came from service with Norwegian and Swedish merchant vessels, the Canadian Coast Guard, and Mediterranean yachts.

Later, I attended Simon Fraser University in British Columbia to study archaeology, linguistics, and communications. I have a passion for history and my studies ranged from learning how to make Paleolithic stone tools to reading Egyptian Hieroglyphics.

In November 1971, I sailed with the Don't Make a Wave Committee to protest nuclear testing by the United States Atomic Energy Commission on the island of Amchitka in the Alaskan Aleutians. In 1972 we officially changed the name of the Don't Make a Wave Committee, and I became a co-founder of the Greenpeace Foundation.

I left Greenpeace in 1977 and founded the Sea Shepherd Conservation Society. Since 1977, I have headed the Sea Shepherd Society and continue to do so. In that time, I have led hundreds of sea going expeditions to oppose illegal activities exploiting marine wildlife species.

During this time, I was responsible for or oversaw the sinking of numerous illegal whaling ships, the sinking of one sealing ship, the blockading of a Canadian sealing fleet, the destruction of a whale processing plant, the paint bombing of a Soviet trawler, the ramming of two Japanese drift net vessels, the ramming of a Mexican tuna seiner, the ramming of a Taiwanese drift netter, the shutting down of a Japanese dolphin hunt, the closing down of seal hunts in Great Britain and Ireland, the landing in Soviet Siberia to document illegal whaling activities, the expulsion of the Japanese whaling fleet from the Southern Ocean Whale Sanctuary, and the interference in killing operations against marine mammals in thirteen countries.

My most famous activities have been the eight voyages I have led to oppose the unlawful whaling activities of the Japanese whaling fleet in the Southern Ocean, the last four campaigns of which were featured on the Animal Planet show Whale Wars.

Most importantly, these things were accomplished without causing or sustaining a single injury to our opposition or to my crew. Additionally, these missions did not result in a single criminal felony conviction against my society, my crew or myself.

Even more importantly, these missions resulted in the salvation of the lives of tens of thousands of whales, hundreds of thousands of dolphins, and millions of seals, not to mention hundreds of wolves and elephants and hundreds of thousands of trees in non marine issues that I championed.

What I learned from Musashi was the importance of mastering numerous arts and skills. For myself this includes writing, poetry, public speaking, debating, martial arts, fencing, go gatsu, navigation, seamanship, and a knowledge of philosophy, religion, geography, science, astronomy, and history (natural, anthropocentric and cultural).

I mention the above accomplishments as proof of my credibility in addressing the issue of strategy.

Therefore, armed with the penned ingenuity of Sun Tzu, the
Self-discipline of Musashi, the perceptions of media and
culture by McLuhan, I can now incorporate my own
personal experience formulating this book.
With the heartfelt and sincere wish
that I may make a contribution
in a small way-towards
directing your travel
along the path of
a Warrior for
and of the
Earth.

FOUNDATION

Strategy is to be found through five paths. These paths are named after the natural spirits of Earth, Water, Fire, Wind, and Dreams.

The Earth is the foundation. It is difficult to realize strategy through a single discipline. Know the smallest things and the biggest things, the shallowest things and the deepest things. Always keep an open mind and realize that no answers are final nor firm. This is the first step to realizing the path of the five ways.

The second way is Water. With water as the basis, the spirit becomes like water. Water adopts the shape of its receptacle, it is sometimes a rivulet and sometimes a wild sea. Water is clarity. Water is purity. Water is calm, Water is violent, Water is aggressive, Water is passive, Water is life. The spirit of victory is the same, be the victory small or great. The strategist makes small things into big things. In principal of strategy, the water path is to have one thing and to know many things.

The third way is Fire. This is the path that deals with fighting. Fire is fierce regardless of the size of the flame. The spirit can become big or small. What is big is easy to perceive. What is small is difficult to perceive. It is difficult for large numbers of people to change position, so their movements are easy to predict. An individual that can easily change movements is difficult to predict. Fire is spontaneous. Fire is comfort. Fire is violent. Fire transforms. The essence of Fire is that you must train day and night if you are to make decisions without hesitation. In strategy it is necessary to regard training as a normal routine of your life without changing your spirit. This is the path of combat and confrontation.

The fourth way is the way of the Wind. This is the understanding of the lessons of the past and of all cultures. It is difficult to know yourself if you do not know others. To all roads there are side paths. You cannot study one way and allow your spirit to diverge. If you are following one way and allow yourself to diverge a little, this will later become a large divergence.

The fifth and last path is the road into the Vision world. This is the lesson of dreams. The Vision quest is that which has no beginning and no end. To be a dreamer is passive but to be a dream warrior is power. Controlling dreams and transforming dreams into reality is the way of the warrior dreamer. Attainment of this principle means to not attain this principle. The way of strategy is the way of nature. When you begin to understand the power of nature, knowing the rhythm of all situations, you will begin to be able to act naturally and to fight naturally, intuitively, without thought or hesitation.

In summation:
The Earth is your foundation. Seek to be aware of all things. Keep an open mind and understand that you cannot and you will not ever perceive the truth that is reality. There are many realities. Hold on to your own reality and evolve from it. Do not dismiss the reality of others.

"The answer is there is no answer."
- Gertrude Stein

Bend like the willow but with the heart of the oak. Flow like water. Be clear, be pure, and cherish your achievements. Build upon your strengths and be creative. Be focused, but do not lose sight of what is around you.

In conflict, you must be fierce. Learn to calculate. Cultivate the skill of Seeing. You must appreciate discipline. You must be prepared. You must be aware of the past. The study of history is imperative. You must be aware of what is happening in the world of the present- all things in general.

However, never lose your focus.

You must be in tune with the natural rhythms of the Earth. You must surrender reason to intuition. You must not fear death.

THE STRATEGIC ARTS

Strategy is the craft of the Earth Warrior. There are few Earth Warriors in the world today who really understand the Way of Strategy.

There are various strategic arts. There is:
        The strategy of the Healer.
        The strategy of the Communicator.
        The strategy of the Artist.

The strategy of the Infiltrator.
The strategy of the Catalyst.
The strategy of the Shaman.

A human must choose a strategy based on intuitive inclination.

The strategy of the Healer is to restore. This may mean restoration of an ecosystem or the practice of natural healing arts as applied to living organisms.

The healer seeks to restore, to rejuvenate, to refresh, and to repair. For the human practitioner, the objective is to counter the negative consequences of human action with restorative actions. This way requires a gentle nature, a caring nature, and the ability to absorb the negative from without while preserving the positive within.

The strategy of the Communicator is to communicate ideas and to teach the Way of all Earth Warriors. This strategy requires a curious nature, the skill of writing, and the ability to speak and to be understood.

The strategy of the Artist is to demonstrate respect for the Earth, to bridge the ever widening gap between natural humanity and alienated humanity, and to connect the present with the past and the future. This strategy requires intuition, imagination, and a natural ability to communicate effectively.

The strategy of the Infiltrator is to affect change from within the fortifications of enemies of the Earth. This strategy requires the skills of espionage, creative sabotage, and proficiency in the skills valued by the forces of anthropocentrism. This Way requires strong determination and dedication, a mastery of the art of deception, absolute confidence and supreme courage.

The Way of the Catalyst is to confront and to provoke. The followers of this are like acupuncture needles stimulating a response. Those who follow this Way are also like lightning bolts, striking quickly and with consequence. This strategy requires great courage, stubbornness, determination, ambition, leadership skills, and complete dedication to the Way of the Earth Warrior.

The strategy of the Shaman is to connect the spiritual reality to the material reality. The Shaman is a spiritual guide for all Earth Warriors. The Way of the Shaman requires the skills of intuition, the ability to tap and be guided by the dream world, and the skills of perception and empathy. The Shaman must be selfless and dedicated completely to the power of the Earthforce.

Whatever strategy is chosen, it must be remembered that the Way of the Earth Warrior requires an absolute acceptance that the Earth is life eternal and individuals must die. The fear of personal death must be overcome in order for an individual to progress as a Warrior serving the Earth.

The Earth Warrior must overcome all concerns of personal welfare and personal ambition. To think only of the practical benefit of knowing the Way is to lose the Way.

When your thinking rises above the concern for your own welfare, wisdom which is independent of thought appears. Whoever thinks deeply on things, even though with careful consideration of the future, will usually think around the basis of personal welfare. The result of such thinking will bring about negative results for the Earth. It is very difficult for most people to rise above thinking of their own welfare.

So when you embark upon something, before you begin, you must fix your intention on the Four Oaths and put selfishness behind you. Then you will not fail.

THE FOUR OATHS

1. Never lose respect for the Way of the Earth Warrior.
2. Be of service to the Earth and all of her children.
3. Be respectful to the Earth and all of her children.
4. Get beyond love and grief, beyond personal ambition.

Exist for the good of the Earth and all of her children. Never be concerned with the idea of success or failure. The Way of the Earth Warrior is to see things through.

Recently there have been some people getting on in the world as environmentalists, as conservationists, as defenders of the rights of animals, as healers of the plights of the Earth and the afflictions of people. If we look closely, we see that these people in many cases are motivated by profit or glory. They have corrupted the strategy. Some men and women are using the issues and taking advantage of the rape of the Earth and her children to sell themselves. They are looking for profit or fame. Some are looking to co-opt the cause to further anthropocentric agendas.

Immature strategy is a cause of grief.

THE STRATEGY OF THE HEALER

To follow this strategy it must be understood that the Earth is a living organism, as are all planets, as are all stars. Life is more complicated than we can ever hope to imagine. The diversity of life is infinite, relative to the finite ability of the human mind to comprehend. We will never have the capacity to understand all. If there is a power that exists and that does understand all, then it is not within the realm of human attainment to ascertain such things. All attempts to do so will result in conflict and failure.

We must learn to be content with not consciously knowing. This is not difficult once you accept that you are a part of all that there is and that you will always be a part of all that there is. There is no collective death, only continual individual death and renewal. Because you are an integral part of the Earth, knowledge will make itself known to you only when it is necessary that you know and only when your intent is clear. The Earth is the collective sum of her parts. The organic and the inorganic chemistry of her being combine with the immaterial to make her whole. Because of this, all of her parts are important and must be respected.

The strategy of the Healer is an understanding that to heal a piece of the Earth, to heal a human, a tree, an elephant, or a river is a task that is equal in respect. The life of a human is equal to the life of an animal. The life of an animal is equal to the life of a plant. The life of a plant is equal to the state of being of a mountain or a lake.

The definition of alive is a state of being. Water is life. Rocks are living. Life can be defined as IS, the state of being. If an Earth Warrior decides to apply the strategy of the Healer, this must be understood. The strategy is the same but the skills and tactics vary.

To heal a human being, the Healer must be knowledgeable of anatomy, of the properties of herbs, the benefits of nutrition, the needs of the body, mind, and spirit. The Healer of humans must know how to set broken bones, to abate a fever, and to direct energies towards helping the body to heal itself. These people may study naturopathic or homeopathic skills or the art of shiatsu or other hands-on skills. They may teach and direct in the fields of physical education, nutrition, psychology, or meditation.

There are many healing skills. The choice is open, the strategy remains the same.

The same strategy applies to the healer of nonhuman animals and differs naught. The skills and knowledge will differ. The anatomy of non-human animals must be studied. An intuitive understanding of the nature of nonhumans is a necessity.

The Healer of non-human animals may become a naturopathic veterinarian or an individual that works for the welfare of these beings. Animal welfare workers may choose to work with domestic or wild animals. Conservationists will seek to protect the rights and habitat of wild animals and protect them from the onslaught of the baser human animals and their domestic non-human slaves. The application of the art of healing animals by humans must be for the welfare and well-being of the animal and not for the purpose of human utilitarian needs.

The same strategy applies to the Healer of plants. Again, the skills and knowledge will differ. The workings of plant-life must be understood. The healer of plants may become a botanist or a gardener. It is important here to remember that for the human healer of plants, the healing must be applied for the good of the plant and not for the purpose of human utilitarian needs.

The same strategy applies to the inorganic. The skills and knowledge will differ. The workings of inorganic chemistry must be understood.

Regardless of the being in need of healing, the Healer must respect the recipient of the healing process. At the same time, the Healer must respect the interdependence of other states of being.

THE STRATEGY OF THE COMMUNICATOR

The art of the Communicator is to teach the nature of the Earth to others and to communicate the laws of ecology. The responsibility for communicating the strategies of the Earth Warrior lie with the Communicator.

If the Communicator takes the path of the teacher, then the primary purpose of teaching is to show the paths of the Earth Warrior. Teach by example. Do not dictate. Teach with respect, both for the art and for the student.

Above all, the teacher must be aware of the responsibility inherent in this path.

If the student is a child, it is important that the teacher understand the nature of children. It is important that the child both trusts and respects the teacher. Respect also the intuitiveness of the child. Nurture the natural abilities of children. It is not so important to teach a child as it is to direct the natural curiosity and intelligence of a child through guidance and support.

If the student is an adult, it is important that the teacher appreciate the dependence of adults. Understand that the adult mind is not as open as the child. Understand the limitations of the adult mind. It is the rare teacher that can overcome the limitations caused by the prior conditioning of adulthood.

If the path of the journalist is chosen, it is important to recognize that objectivity in media is rare and essentially unnatural in human cultures. Journalists favour either the right or the left, the biocentric or the anthropocentric. The journalist who sides with the Earth must utilize the media as a tool to promote the best interests of the Earth.

The art of the journalist is important because the media is the key to reaching the hearts and minds of the general population. This discipline works best when coupled with the strategy of the infiltrator. However, the strategy can be applied overtly in the limited media of the movement.

THE STRATEGY OF THE ARTIST

The strategy of the Artist is a difficult discipline to follow. This strategy requires self-discipline both in applying and mastering a natural skill and in resisting temptation from the material world. When following this way it is important to never forget that the objective of the artist is to serve the Earth, to bridge the ever widening gap between biocentric humanity and humanity alienated from nature.

This strategy requires a powerful intuition and an even more powerful imagination.

The strategy of the Artist has in recent years been pre-empted by the material world with co-opting and indulgence of the ego. Thus the most important strength to cultivate for the follower of this strategy is humility.

The best of artists have a gift for perceiving other realities. It is important that the artist serve the Earth and the future and not just self interest. It is important that the artist serve the Earth and not the pseudo world of art itself.

For many people, art has become a paramount concern. Some people fervently believe that the works of humanity are of utmost importance. The pyramids, the Old Masters, the symphonies, sculpture, architecture, film, photography, and the things that we have chosen to define our culture or to define ourselves create a belief that in these fleeting material things we will find our identity. All of these things are worthless to the Earth when compared with any one species of bird, or insect, or plant. The creations of the Earth are supreme and the only importance. The creations of humans are primarily vanity.

Selling a Van Gogh painting for sixty million dollars is an obscenity. The artist did not profit by this sale. This particular Japanese buyer raped thousands of square miles of rainforests from all over the world to acquire the wealth to purchase a bit of coloured hydro-carbon splattered on canvas. The artist was debased by the greedy destruction of biological complexity.

The strategy of the Artist must recognize that works are merely reflections on the natural beauty of the Earth and Universe.

People admire, acquire, and possess artworks in the hopeful belief that in these human generated things they may find their centre, their identity. Very few do. Only those who actually practice the strategy of the Artist find their center in the act of creating. There is virtue in admiration and appreciation of art of course but art works best when it inspires and conveys a powerful message both to the heart and the mind.

It is important that the artist contribute to the welfare of each subject. If a painter specializes in whales and if the work is profitable, then the profit should benefit the whales. If a musician is inspired by the trees or the mountains, then contributions should be given for the protection of trees and mountains. If the artist follows the path of architecture, it is important that his or her creations do not embarrass or conflict with the Earth.

THE STRATEGY OF THE INFILTRATOR

This is the art of espionage and covert sabotage. The Infiltrator should not have a public record of actions in support of the Earth. This type of Warrior must be able to assimilate culturally with the opposition.

Overtly, a practicing member of an anthropocentric religion, covertly a dedicated follower of the natural spiritual realities.

Overtly, a person who respects the anthropocentric laws, covertly a dedicated follower of the natural laws of ecology.

Overtly, a person who respects the anthropocentric system, covertly a person dedicated to opposing the forces of anthropocentrism.

This is a difficult Art. The temptations are great. It can be a lonely position.

It is also one of the most effective means of opposing the opposition.

It is wise for the Infiltrator to liaison only with a person who follows the strategy of the Catalyst. It is wiser still to liaison with no other person.

This is a path without recognition or of belated recognition, sometimes posthumous recognition, sometimes with the stigma of being historically cast as an enemy of the Earth. And yet, human perceptions of your contributions to the Earth and the children of the future are not relevant.

The Infiltrator should be well educated within the confines of the established educational system.

The Infiltrator should never give overt gifts to groups following the strategies of the Earth Warrior. This can be traced.

Infiltrators may choose careers in law, in business, in the military, in government, or in the diplomatic service.

Infiltrators can covertly pass on information to concerned groups of individuals.

Other tactics that can be utilized are: the spreading of disinformation, outright sabotage of technical equipment and plans, erasing files to protect people in the Movement, and covert re-direction of funds.

Sometimes an infiltrator may choose to go public if the advantages are worth the cost of exposure. This would mean directing yourself towards the strategy of the Catalyst.

Sometimes an infiltrator may be overtly sympathetic but is beyond direct proof of involvement.

This is the primary strategy of the intelligence agent and the foundation of effective espionage.

THE STRATEGY OF THE CATALYST

The art of the Catalyst is a demanding discipline. The main objective of this strategy is to act like an acupuncture needle to stimulate actions and change.

This type of Earth Warrior must maintain a high public profile. Experience in manipulating media is important. The individual must be a good speaker, a good writer, and a person of action.

This is a strategy where the Warrior must lead by example. Charisma is a quality that should be cultivated. The Catalyst must be strong, stubborn, righteous, and maintain a positive attitude.

The Warrior choosing this strategy will lead, establish, and direct organizations.

In the Movement, those who follow the strategy of the Catalyst will be public speakers, media spokespeople, and leaders of actions.

Catalysts must continually expose themselves to danger, to public ridicule, to the equally annoying public adoration, and to harassment by the guardians of established law, the police, and intelligence agencies.

The strategy of the Catalyst is followed by the generals of the Movement.

They are the primary strategists for the directions of operations of a direct nature.

THE STRATEGY OF THE SHAMAN

The Art of the Shaman requires rigid self-discipline. The Shaman is the connection between the spiritual interpretation of the forces of nature and those warriors of the other paths who have also chosen to serve the Earth.

The true Shaman does not wear the cloak of a guru. The virtues here are wisdom, intuition, humility, discipline, and sacrifice.

ALL ARE ONE

The Shaman is the Light of strategy. The Healer is the Heart of strategy. The Communicator is the Voice of strategy. The Artist is the soul of strategy. The Infiltrator is the Eyes and Ears of strategy. The Catalyst is the Will of strategy.

Together the disciplines harmonize into a working movement of individual warriors united by a bond of love, respect, and duty to the Earth.

EARTHFORCE!

The Spiritual Foundation of an Earth Warrior

A Warrior's life is an eternal quest for wisdom, harmony, and understanding. It must, however, be appreciated that complete knowledge, peace, and enlightenment will not be realized. If you do not accept this, then disappointment, despair, and cynicism will overwhelm you in the future.

It is the quest that is real and it is in the quest that near perfection, near harmony and a glimpse of understanding may be found.

For yourself, there is only now - the present. The past belongs to those who came before. The future belongs to the children. Lessons can be drawn from the past. Concern and love for those unborn may guide your actions. But it is only the present that you can bend to your will, only the present that sustains you, and it is only in the present that you can act. The past is for scholars. The future is for dreamers. The present is for the true activist. Although it must be said that a true Warrior is a scholar, an activist and a dreamer.

The most important trait for a Warrior to cultivate is self-discipline. In fact it is only self-discipline that separates a Warrior from those that simply exist.

You must rise above the masses of humanity; through sacrifice, by force of your own will, you can go where others dare go, and you will see what others fear to look upon. You will experience life in the way it is meant to be - experienced to its fullest.

It is not my place to tell any other how to think or what to think. Your quest is your own. Your way must be your way, not mine, and certainly not the way of the collective mass of humanity, the destroyers of nature and the befoulers of the Earth.

I am simply a guide and a humble one at that. I can only direct you towards the path with the benefit of the little knowledge that I have acquired. I cannot walk the path with you nor can any other person. You must walk it alone with courage and with dignity.

But I can tell you that to be successful you must attempt to clean the slate of your heart of all the nonsense that anthropocentric society has stuffed into your head. You must throw off reason and cultivate intuition. You must allow your guidance to come from within and no longer from without.

You must touch the Earth. It is from the Earth that you have sprung and it is from the Earth that you receive your nourishment and your strength. The Earth Warrior serves the Earth. The Earth comes first - always.

The protection, conservation, and preservation of the Earth is the foremost concern of the Earth Warrior. You must be prepared to risk all, including your life and liberty, to uphold the sacred integrity of the Earth.

You can do this only if you truly believe in the sacredness of the Earth, the sanctity of nature, and the holiness of the wilderness. The major failing within the environmental movement is the failure to see the Earth as sacred.

Anthropocentric culture has taught most of us to look upon anthropocentric beliefs as sacred. Thus it is considered blasphemy to spit upon the Black Stone in Mecca or to dismantle the Wailing Wall in Jerusalem or to desecrate a statue in the Vatican. If any person were to do any of these things, they would be dealt with quickly and violently and anthropocentric society would applaud their punishment. Yet when loggers assault the sacredness of the remaining redwood forests of California, when they desecrate the cathedrals of the natural world, the environmental movement responds with petitions and letters or protest signs.

If our view of the redwood forests is that they are sacred, then their destruction should be considered by us to be blasphemous, and the destroyers should be quickly and violently dealt with. To many environmentalists, the talk of sacredness in nature is nothing more than rhetoric. The desecration of the sacred, if the sacred is really sacred, will always inspire gut anger and will in all cases provoke an active response.

To the Earth Warrior, a redwood is more sacred than a religious icon, a species of bird or butterfly is of more value than the crown jewels of a nation, and the survival of a species of cacti is more important than the survival of monuments to human conceit like the pyramids.

The natural rage inspired by those who violate or attack the sacred must be channeled by the Earth Warrior through discipline. The enemies of the Earth can only be met with an opposition that wields superior strategy and superior tactics. Numbers and technology can always be defeated with the application of superior strategy and tactics.

In anthropocentric society, a harsh judgment is given to those that destroy or seek to destroy the creations of humanity. Monkey-wrench a bulldozer and they will call you a vandal. Spike a tree and they will call you a terrorist. Liberate a coyote from a trap and they will call you a thief. Yet if a human destroys the wonders of creation, the beauty of the natural world, then anthropocentric society calls such people loggers, miners, developers, engineers and businessmen.

A strong positive spirituality must be cultivated with its roots firmly implanted in a biocentric perspective.

The Earth Warrior is a biocentrist. An Earth Warrior serves the biosphere.

For an Earth Warrior there is no compromise with the anthropocentrist.

The Earth and her children are first. The interests of the whole is more important than the self interest of one species, any species including our own.

The Earth Warrior is equal to the wolf, the whale, the willow, the wasp, and the wren.

Although it must be stated that some species are indeed more important than others. Worms, ants, bacteria and fish are more important than human beings for the simple reason that worms, ants, bacteria and fish do not need humanity to survive but humanity needs them to survive. They are essential for our existence. We are not essential to their existence.

Species are more important than individuals.

Within the context of species equality can be found a sense of planetary belonging. To be a part of the whole is to be free of the alienation caused by an individual species like our own becoming divorced from the biospheric family of life.

To be spiritually liberated from the chains of anthropocentrism, you must ruthlessly abandon all attitudes that place humanity on a special pedestal of worthiness.

Christ, Mohammed, Buddha and Ron L. Hubbard were primates, cousins to the chimpanzee and the mountain gorilla, just as we are all the children of a species of naked simians. In primateness, we can find family. Although anthropocentric humans may consider themselves to be divine legends in their own mind, the biological reality is that they are simply overly glorified, conceited apes.

God is not a monkey. God is the unknown, the feminine force in divine harmony with the masculine force, the is and the is not, the positive and the negative, the alpha and the omega. God is anything and everything that can be imagined, But the one thing God is not is definable. Attempts by the anthropocentrists to define the unknown, to finite the infinite, have only led and continue to lead to ecological holocaust and the deaths of billions of lives.

All wars have their foundation in the conflicts of anthropocentrism over philosophy, territory, resources and power.

In embracing the biocentric foundations of a natural biocentric perspective, you must abandon and dismiss the dictates of anthropocentric theologies.

The truth is not to be found in Christian icons, Judaic writings, Islamic rules, Buddhist meditation, dianetics or any of the theological inventions of humankind.

The truth cannot be found in political philosophies.

The truth can only be found by realizing that it cannot be found, and that all is not knowable, that the infinite can never be comprehended by the finite. The purpose of life is life itself. The purpose of death is rebirth. The purpose of birth is death. The immortality of humanity as in all species is to be found in the Continuum.

In the ballads of the Frankish knights lie encoded the true essence of the warrior's path. The Grail romances are the foundation of European chivalry, in King Arthur we find the definition of glory and nobility which is great valor in a cause of certain defeat. Arthur's greatness was in serving a cause greater than himself, knowing that he would inevitably fail. In his failure he won for his cause the time needed by those he did serve.

Environmentalists serve the Earth and we fight to buy time and preserve territory, a minute at a time, an inch at a time. Our victory lies in accumulation of effort. To serve the cause effectively lies in the ability to direct passions most strategically over the long period of time.

THE CONTINUUM

Time flows like blood through the veins of reality. Time is the measure of all things. To all living beings, time dictates a beginning of self-awareness and an end to self-awareness. To exist within the flow of time is to take on the reality of sentience for a brief period before returning to the reality of nonsentience.

Life is energy and energy cannot be created nor destroyed. Life is a transition, a link between the sentience of the past and the sentience of the future.

All living things upon the Earth are a part of the Earth. All living things upon the Earth have always been apart of the Earth. All living things upon the Earth will forever be a part of the Earth. For in truth, all that is alive upon the Earth IS the Earth herself. All is one. We are the Earth and we are indivisible from ourselves. All life is interdependent.

All that came before and all that will come later are also one and the same.

Past, present, and future are different stretches of the same river. Like the molecules of water in a river, the living beings of the past remain connected to the living beings of the future through the living beings of the present.

This is the Continuum of life.

At the present time, humanity is divided unequally into two distinct and radically different perspectives. People are anthropocentric or they are biocentric.

Anthropocentric people have chosen to step out of the flow for a brief period. They no longer feel the connection to their ancestors nor do they feel kinship with the children of the future. They do not even feel kinship to the other non-human beings of the present.

I said for a brief period and this is because anthropocentric thought which is a deviation, will exist only briefly before self-destructing, The Continuum is the guide for navigating the river of life. Without the Continuum, life has no direction and such a life-style will run counter to the Ecological Laws and thus will create ecological disorder.

The biocentric person naturally takes an interest in the people of the past.

Aboriginal people are well aware of the deeds and lives of their forebears, going back hundreds of years. Anthropocentric people give little thought to the deeds and lives of their grandparents or even that of their parents.

Children of all cultures express an interest in those who came before them. The people of the Earth will share their knowledge of their ancestors with their children; The people who have divorced themselves from the Earth will dismiss this knowledge as unimportant.

A knowledge of history, both natural and human, is an essential discipline for the Earth Warrior. Natural history gives us an understanding of evolution and thus a more humble appreciation of our rightful place as human animals upon this planet.

Natural history teaches us the connection between all living beings of the present with all living beings from the past. Natural history is the story of the Continuum of all life on Earth.

The written history of humanity is an anthropocentric history for the most part. There are, however, glimpses of wiser cultures within the written record, although these cultures have usually been recorded only when they have hindered or have been a victim of anthropocentric progress. Records of pre Columbian people of Turtle Island now called the Americas or pre-European cultures of what is now called Australia present fascinating glimpses of biocentric cultures. The same holds true for early pre-Roman Northern European history or pre-Bantu South African history.

From human recorded history we can learn about biocentric and anthropocentric attitudes and the evolution of the two, the conflict between them and the comparative ecological impact of both. It is through history and anthropology that we can learn from the mistakes of our ancestors. It is through history that we can resurrect the ecologically more positive life styles of our more enlightened ancestors. It is through history that we can learn the results, both success and failure of the application of different strategies.

The study of history also expands our knowledge of events past so as to draw us into the reality of the past. This will give us a more solid position within the Continuum of Life. To not have an understanding of history is to be like a person suffering amnesia. Without memory you are not complete. A memory of one's personal life may be sufficient for the anthropocentric individual. The biocentric person, on the other hand, needs the knowledge of the past to be fulfilled. Without this knowledge, the past within the Continuum is foggy and perceptions of the present are distorted.

We can approach our past in the Continuum through knowledge of history. An understanding of the future can be had by cultivating our natural, intuitive ability to envision.

Vision is a difficult ability to harness if stifled by rational or logical thought. Vision comes from the heart and soul and not from the head. Vision is the ability to see the past, the present and the future.

Through knowledge and intuition we can see and feel and thus understand the past. Through intuition and our senses we can perceive the present. Through intuition and imagination we can envision or dream the future.

Most anthropocentrics have little desire to glimpse or even to be curious of the future. The average anthropocentric person looks into future concerns, the next pay cheque, the next mortgage payment or the possibility of sending their children to a suitable university. Their idea of the future is self-interested. When they do talk of the more distant future, they usually limit their discussions to how they see the world to come in terms of technological innovation. Talk of space travel, of more efficient timesaving gadgets, or life extension discoveries monopolize most of the discussions. Very little is said about pollution, diminishment, species extinction, over-population, scarcity of food and water, and on and on. A blind faith in human technological inventiveness allays all concerns in this direction. To the anthropocentric mind, tomorrow will of course be better than today because they also believe that today is better than yesterday.

This vision and the reality our descendents experience may be radically different.

To the biocentric, the children and people of tomorrow are real people. The children of the world of five hundred years from now are as much our children as those who directly and physically came from contact with our own bodies.

The biocentric loves not only the child of his or her loins but also the baby child and adult who is birthed by the woman who will be a reality by virtue of a direct link between the now and tomorrow.

Those who call themselves deep ecologists today work most directly for the welfare and well-being of all those who will live throughout the generations to come. Environmental activists today may be reviled, ridiculed, hated, and condemned by anthropocentric attitudes, but to the people of tomorrow (if we survive) they will make excellent ancestors.

Environmental activists represent the majority of humans because we represent all those billions of people who have yet to be born over the next ten thousand plus years. In addition, environmentalists represent the billions of individuals of the tens of millions of fellow citizen species.

Born of the Earth we return to the Earth. The soil beneath our feet contains the material reality of the ancestors of all species. Without the expended lives of the past, there would be little soil. For this reason, the soil itself is our collective ancestry and thus the soil should be as sacred to humans as it is to our fellow citizen species.

The water of the Earth is the blood of the Earth and within its immensity will be found the molecules of water which once enlivened the cells of our ancestors of all species. The water you drink once coursed through the blood of the dinosaurs, or was drunk by Pre-Cambrian ferns or was expended in the urine of a mastodon. Water has utilized the lives of all living things as a part of its circulatory system. All life contains water. Water circulates through the earth and atmosphere, through glaciers and in the bodies of all living organisms. Water is the bond that holds all life together and connects all species intimately forever. Therefore water is sacred.

The air that we breathe has passed through countless living respiratory systems and thus has been chemically stabilized by plants and animals. Without the lives that have gone before there would be no air to breathe. The gases of life emerge from the breath of all living organisms. The life of all past organisms have nurtured the atmosphere. Therefore the air is sacred.

Because soil, water, and air are sacred due to the contribution of all past lives, then it must be recognized that the past is sacred, and thus the memories of our ancestors are sacred.

Our lives in the present should be sacred to all the living things of the future. In the realm of humanity however, the contribution of humans of the present will be as an object lesson on why natural human beings should not stray blindly from the path of nature and outside the flow of the Continuum. If our species survive we will be known as The Lesson.

For those in the present, the only way to contemplate the future is through feelings of love. Love for our children and our children's children and their children ad infinitum or to the end of our species and beyond to the end of the Earth.

Our actions and our decisions must always weigh the consequences for future peoples. There is an Iroquois expression that states, "make no decision without first taking into account the impact that it will have on all future generations."

When you father or mother a child, you are obligated within the Continuum to bear responsibility for the protection, nourishment, nurturing, and education of the child. To not do so is the greatest of irresponsibilities and one of the gravest crimes a human can commit.

With the birth of a child, the parents set in motion a series of events which will ripple forward into the future of the Earth. With the birth of one, you have the potential to bring forth thousands. The psychology of one generation deeply affects the psychology of the next and so on.

Within the context of nature, all animals abort their young if they cannot guarantee the welfare and health of their off-spring. In some cases, the very fact of an increase in population can jeopardize the health of the child. In such a case, the unborn IS sacrificed in order to maintain the stabilization of the natural order.

Abortion is a serious responsibility and it is a responsibility that must be borne by the females of all animal species. Males and females have responsibilities equal in value to the welfare of both. Abortion however is a female responsibility. Only she can decide if she is psychologically and physiologically capable of giving birth and sustaining the development of the child. For this she must also be secure in the support of her partner. The decision is hers and the emotional burden is also hers to bear and hers to deal with, again hopefully with the support of the responsible partner.

As long as the unborn baby is physically connected to the mother, the child and the mother are one and the same. The child is not an individual if it cannot sustain itself. The woman, as mistress of her own body, has every right to judge and decide on the welfare and future of her body, mind, and soul.

When anthropocentric society dictates that the law of the state over-rules the freedom of the individual to decide her own fate and chart her own course, such legislation amounts to the rape of the individual woman by the brutish might of the faceless state.

To force a woman to produce a child against her will is rape.

By outlawing pre-natal abortion, the state brings into being the practice of post-natal abortion and thus a child is set on the road of life burdened with the emotional scars of being unloved, uncared for, and abused. Deviant criminal behaviour, from sexual crimes to mass murder, can be attributed to the ranks of those who had post-natal abortion inflicted upon them by the dictates of the nation state over the responsible choice of the Mother.

This is not to say that all those who decide to have children are making responsible choices. So alienated from nature have we become that we as a species have been grossly guilty of bringing forth children without the guidance of the Continuum.

As a result, the most dangerous threat to the future has come about: the lethal explosion of human populations.

Anthropocentric society values quantity of life over quality of life. The greater the numbers of humans on the planet, the cheaper each individual life becomes. As a result, anthropocentric society has slaughtered over one hundred million people in the twentieth century alone in senseless wars over territory and ideology.

Nation states require population growth, if not through births then through immigration. Numbers mean political strength and bodies to employ in industry, agriculture, and warfare. These are the three great plagues that anthropocentric society has inflicted upon the natural world.

The Catholic church condemns both birth control and abortion because its institutionalized strength lies in the number of worshippers and not in individual devoutness.

In a biocentric world like the one that humans lived in for tens of thousands of years before the invention of agriculture, small groups of humans can live harmoniously with nature. Today, evidence of this harmonious relationship can be seen in the cultures of the bushmen in Africa or some of the native tribes in South America or Indo-China.

Prior to the development of agriculture, the human species stabilized itself with zero population growth. Since the development of agriculture and especially since the development of industry and technology, the Earth has witnessed incredible exponential human population growth. In fact, there are now more human beings alive on the planet in the present than have cumulatively existed in the past before 1800.

The only solution to this situation is for the implementation of negative population growth. Since negative population growth must ethically be through choice, the outlook of achieving it within the context of an anthropocentric culture is remote.

But unless the human species makes such a choice to consciously and humanely lower its population, then we must anticipate the intervention by the laws of ecology and the consequences that this will bring.

Our deviant behaviour will be corrected by nature when our populations exceed the carrying capacity of the Earth. This will happen within the next century. The calculations based on present day population growth indicates a population of nearly twenty billion by 2100 A.D. Once our populations exceed carrying capacity, the population will crash due to natural factors like starvation, disease, lack of water, lack of shelter, pollution, and conflict. Prior to the collapse there will be an insidious decline in the quality of life for all living things.

This crash has already begun in many parts of the world and will rapidly escalate.

There will be survivors just as there were survivors of the great killer plagues of Europe which began in the twelfth century. However, the experience will not be a pleasant one. The death, the misery, the suffering, and the humiliation of such an experience will make all the trials and tribulations of past human history pale in significance and impact.

And the survivors, unless they return to the Continuum, will set in motion a pattern of life styles which will lead again and again to population crashes.

Such a legacy should not be bestowed upon future people by our lack of vision today. Do we wish to leave our children's children with a heritage of death, misery, and destruction or do we wish to transmit our love through the ages by taking steps to preserve the beauty and diversity of the living Earth as it is today, as it has survived to this day?

Remember, we, the children of today, were robbed by many of those of our ancestors who slaughtered, raped, and pillaged their way through the natural world. There are thousands of animals and plants that did not survive because of the lack of vision expressed in the greed of some of our forebears. Today, the greed of many of us will rob the children of the future of similar treasures.

It was only during the last five hundred years that the great biocentric cultures of Australia, Africa and the Americas were destroyed. In that brief time, anthropocentric human civilization has all but destroyed the natural wonders and treasures of the virgin expanses of lands that were dwelling in natural harmony.

What our children's children need from us is the expression of our rage as Warriors. We need to fight, to set an example by our actions. We need to exert with all of our individual and collective strength the will to place ourselves back within the embrace of the bosom of the Earth.

We need to return to the garden of the natural world. We need to revolt against anthropocentric thought. We owe such a revolution to those of the future. Through our revolution we can absolve ourselves of the sins of the past.

We of the present are responsible as a species for the actions of our species in the past. We of the present are responsible as a species for the well-being of those of our species yet to be born. We of the present are also responsible for our actions in the present and the consequences that these actions have both on ourselves and on other members of our species that share the present with us. We of the present are also responsible for our actions in the present and the consequences of these actions on all living things that share the present with us.

> If we take care to do good in the present,
> the future will take care of itself.
> Past, present, and future.
> Together as one is ...
> The Continuum.

Chapter 1

PREPARATIONS

Animis Opibusque Parati *
Semper Paratus **

Strategy is of paramount importance to the ecological activist and to the environmental movement. Strategy is what decides the outcome of an action, a campaign, or an initiative. It is the superiority of a strategy which determines success over failure, freedom over incarceration, or life over death.

It is essential that all ecological activists be schooled in the art of strategy. Failure to know strategy is irresponsible. Failure to understand strategy may result in injury, imprisonment, or death, to yourself, your followers, or your allies. Worse, failure to comprehend strategy can lead to failure and thus further destruction to the environment and/or its inhabitants.

It would be better for the environment and the environmental movement if a person did nothing rather than attempt to carry out a campaign with no strategy, an incomplete strategy, or an inferior strategy.

There have been numerous arrests and convictions of activists since the first edition of this book. Many of these convictions could have been avoided if the activists had studied the rules of strategy outlined in these pages.

All strategy requires the careful and meticulous laying of plans. All strategy is governed by five constant factors, all of which need to be taken into account and studied thoroughly.

Sun Tzu described these factors as Moral Law, Environment, Situation, Command, and Method and Discipline.

The **MORAL LAW** means that your cause must be honest and just. You must be morally superior to your enemy. This is often called taking the moral high ground.

* Latin: Animis opibusque parati means, "prepared in minds and resources."
** Semper paratus means, "always ready." Semper paratus is the motto of the U.S. Coast Guard.

If you have and can hold the moral high ground, your followers will follow and your allies will support you, regardless of the risks, and undaunted by danger.

In the environmental movement there are two kinds of moral advantages to take into consideration. For your own followers, those who possess the biocentric outlook of the Earth Warrior, it is important that you remain virtuous in your adherence to biocentric principles. As long as your followers are satisfied with the consistent nature of your philosophical foundation, it is then irrelevant how your views are manifested within the strategy projected.

You have two distinct minds. Use both. Your worldly mind will understand the justification for deception in the presentation of your arguments. Your universal mind, the real you, can remain at peace with the knowledge that your deception has only served to deceive your enemies and not yourself or your allies.

Deception is a weapon. Do not direct it towards your friends and, most importantly, beware of self-deception.

Because we are working within an anthropocentric society, we must be aware of the anthropocentric moral high ground. We must be prepared to present a higher anthropocentric moral stance that makes sense to our opponents, but is in fact, more biocentric in nature.

When opportune, you must appear to be more Christian than the Christians without being Christian, more humanistic than the humanists without being a humanist, more patriotic than the patriots, without being a patriot, etc. This involves being thoroughly schooled in anthropocentric, philosophical thoughts and beliefs. All biocentrists must strive to attain a complete understanding of anthropocentric thought. A good strategist must be able to favorably use the opposition's philosophical arguments.

> *"I will never be out Christianed or out Texaned ag'in"*
> - **George W. Bush**
> (On losing an election in Texas and discovering the power of deception)

Praise the Lord and wrap yourself in the flag if this is what it takes to get your message across. It worked for George Bush. It is the strategy that Sarah Palin utilizes so it can just as easily work for us. Strategy is a discipline independent from objectives.

> *"Men willingly believe what they wish."*
> **- Julius Caesar,** *De Bello Gallico*

Environmentalists often assume that they are right and that environmentalism is inherently just. It must be remembered however, that justice is a matter of perception. People see what they want to see and hear what they want to hear. What is just and right in the mind of an individual or group may not be just or right in the opinions of others or the general opinion of the public as represented by politicians and influenced by the media.

> *"Print technology created the public.*
> *Electric technology created the mass.*
> *The public consists of separate individuals,*
> *walking around with separate fixed points of view.*
> *The new technology demands that we abandon*
> *the luxury of this posture, this fragmentary outlook."*
> **- Marshall McLuhan**

In other words the new technology is one of absolute conformity. Modern media is like the aliens in the Invasion of the Body Snatchers. It is like the Borg from Star Trek and "Resistance is futile, you will be assimilated."

Unless!

You need to remove yourself from being manipulated by the media. To survive you need to manipulate media instead of being manipulated by it. Believe nothing at face value of what you read in newspapers or online and believe nothing at face value you hear on the radio or hear and see on television. Analyze everything because the media is the real equivalent to the matrix.

Truth is defined by the media and your truths can be rendered into lies by the power of modern media.

To an Earth Warrior, the biocentric attitudes may be both right and just. To anthropocentric society at large however, biocentrism is an aberration. You must package your moral superiority in an attractive anthropocentric wrapper. Make the media work for you to deliver your message.

Within the environmental and especially the animal rights movement, cults of superiority have been established. Vegan proselytizing and pagan mother Goddess worship may serve to unify small mutual interest cliques, but tends to be very alienating to the populace as a whole. In other words, you should not confuse your projected strategy with rhetoric on witchcraft, paganism, anarchism, new age philosophy, and other minority points of view. Nor should you alienate the anthropocentric public by dressing and behaving in a manner which threatens the majority morality.

The best equipped activist is the one not noticed, the one that infiltrates, pretends to assimilate and thus becomes highly effective.

I am not dictating here that you must dress and conform to the values of the anthropocentric establishment. I am simply pointing out the reality of the consequences should you not do so. When preaching to the Romans, it pays to look and behave like a Roman. People distrust, dislike, and often detest those who appear to be different.

Become a blank canvas that you can paint an identity upon. Do not carry identifying marks or broadcast your beliefs UNLESS it is incorporated into your overall long term strategy.

If you judge an audience to be receptive to a biocentric approach, then proceed. Or, if you judge that a biocentric approach may shock your audience in a positive way in order to provoke thought, then do proceed. An educated audience will be more receptive to biocentrism than an uneducated audience.

If you proceed in the advancement of biocentric thought then you must be able to explain your position with knowledge and with spirit. An intellectual understanding of biocentrism is insufficient. You must intuitively understand human existence within the context of the biosphere. You must understand the principles of ecology and of natural laws. You must be able to feel the rhythm of the seasons, sense the phases of the moon, and be attuned to the forces emanating from the Earth and from Life.

It is not enough to know that your cause is just. You must convince the public that it is more just than your opponent's cause. The other side will attempt to espouse the morality of their position by citing the need for jobs, tradition, cultural values, or by appealing to a community and creating a "we" versus "them" attitude. Your opposition will attempt to smear you with derogatory labels. You must be prepared for these attacks and counter the public relations machinery and stratagems of your opponent.

Where possible, turn the strategy of your opponent back on your opponent. If a developer uses the promise of jobs to obtain leverage, then point out the health hazards posed to children if relevant, or point out that the jobs are just short term and taxes will rise to pay for the development of utilities and infrastructure.

Never appear too serious, but at the same time be thoughtful. Use wit and humor but never appear too humorous. Do not be cynical. Never speak directly to your opponent, but use your conversation or debate to address the public at large. At the same time, do not appear to be discourteous to your opponent. Never lose your temper in public. Be calm, project confidence, and never be at a loss for an answer.

The nature of the mass media today is such that the truth is irrelevant. What is true and what is right to the general public is what is defined as true and right by the mass media. Ronald Reagan understood that the facts are not relevant. The media reported what he said as fact. Follow-up investigation was "old news". A headline comment in Monday's newspaper far outweighs the revelation of inaccuracy revealed in a small box inside the paper on Tuesday or Wednesday.

> *"All media work us over completely. They are so persuasive in their personal, political, economic, aesthetic, psychological, moral, ethical, and social consequences that they leave no part of us untouched, unaffected, unaltered. The medium is the massage. Any understanding of social and cultural change is impossible without a knowledge of the way media works as environments."*
> **- Marshall McLuhan**

When Ronald Reagan was President of the United States he always had an answer and those answers were not always accurate or even truthful, but he is not remembered for his inaccuracy. He is remembered as the "Great Communicator" because he was never at a loss for an answer and he was able to explain things in the way the average citizen could both understand and appreciate. Don't worry about accuracy and deliver the information confidently and without hesitation. As long as the spirit of your message is truthful, the statistics and the details are secondary. Do not allow yourself to be sidetracked by trivialities.

Since the first publication of this book in 1993, the above statement has been used many times to discredit me. However I stand by the statement. Corporations and politicians lie consistently to the media and the public. President George W. Bush began a war based on lies. This is not a book on the morality of telling the truth, it is a book on using strategy to advance a cause and to win battles and Sun Tzu was very clear in stating that deception is the foundation of strategy.

Without deception there is a weakness and a refusal to utilize deception within strategy can lead to defeat and a lost cause.

For this edition I have decided to expand on the subject of deception with an added chapter entitled **The Disciplined Art of Strategic Deception**.

You should also concentrate on creating symbols for an issue and appeal to the public through these symbols. Powerful symbols are human babies, children, baby animals, beautiful animals, trees, plants, or landscapes. Emphasize the noble, the beautiful, and the innocent. Do not be abstract. The modern media abhors abstractions.

KEEP IT SIMPLE!

If the issue is polluted water or pesticides, utilize the image of children whose health will be threatened. If the issue is wildlife, focus on the cute, the noble, or the beautiful - appeal to anthropocentric preconceived notions. If the issue is habitat, choose an animal that represents that habitat and accentuate its beauty and nobility.

As an example, if the issue is Central American rainforests, then employ the image of the jaguar in your campaign. In Antarctica, use the penguin; in Africa, the elephant. Remember, you will have little success in mobilizing public concern for the mugwort, the sea cucumber, the periwinkle, or a rat. Subtlety attach the cause of the less appealing to the public campaign to protect the more appealing.

Saving tuna is as important as saving dolphins, but dolphins are more appealing. A campaign to protect children in Japan or the Danish Faeroe Islands from mercury poisoning will automatically raise public awareness about dolphins and pilot whales. Both species benefit from a campaign to protect human children.

Be prepared to recruit celebrities to speak for and support your cause. Celebrities are noticed by the public and can effectively influence the public.

Because Sea Shepherd has supporters like William Shatner, Pierce Brosnan, Christian Bale and Richard Dean Anderson, there is the impression that we are a strong organization based solely on the perception of the mythical attributes of Captain James T. Kirk, James Bond, Batman and MacGyver.

Celebrity equates to credibility.

Never worry about ridicule or criticism. Nothing is ever accomplished without enduring such attacks. They are irrelevant and trivial. Confidence is the defence against ridicule and criticism.

Always prepare an effective media strategy.

Remember to always play the media by utilizing the rules established by the mass media.

Remember that the media is not interested in the content of your news. The media is interested in the delivery. Drama, not data. Scandal and sex, not statistics. Violence and visuals, not viewpoint. Despite media pretensions, facts are not a priority for most media. Remember the Fox Network is "fair and balanced."

> *"Societies have always been shaped more by the nature of the media by which men communicate than by the content of the communication."*
> **- Marshall McLuhan**

Never acknowledge a public relations strategy aimed at you, this simply gives increased credibility to your opponent. You must try to find a way to counter any strategy that your enemy puts forth by devising a superior strategy.

Always appear before the media as confident, knowledgeable, in control, and in command. Never debate defensively in the media or in public. Strive to be the David to your opponent's Goliath but without demonstrating weakness. There is a moral advantage to being the underdog. Never appear too powerful, too wealthy, or too strong. Keep any power, wealth, or strength hidden and in reserve. Bait the opponent in, and if you still cannot achieve victory, then unleash your reserves.

Modern media can serve two purposes. First, to increase awareness and second to make things happen. A media strategy can be designed for either purpose or for both.

In the face of negative publicity, make it work for you. All publicity can be turned into positive publicity. As Oscar Wilde once remarked, "The only thing worse than being talked about, is not being talked about."

Use humour at every opportunity. Humour is the best strategy for deflecting criticism.

> *"It is said the warrior's way is the twofold way Of pen and sword."*
> **- Miyamoto Musashi, 1648.**

> *"The way of the Earth warrior is the twofold way Of camera and confrontation."*
> **- Captain Paul Watson, 1984.**

ENVIRONMENT means you must incorporate into your planning the following: night and day, cold and heat, times and seasons.

Always consider proper clothing, equipment, and means of transportation. Do not be caught unprepared for climate or a lack of visibility. Keep a close account of time and strive to be punctual.

> *"The bag's not for what I take, Colson - it's for what I find along the way."*
> **- MacGyver:** *Pilot*

Avoid adverse conditions when defending. Exploit adverse conditions when attacking.

Calculate the advantages of fog or darkness for cloaking your approach. Be aware of the phases of the moon. Always approach out of the sun. Be prepared for parasites, poisonous insects, reptiles, and potentially dangerous animals and plants. Respect these citizens of the wilderness and seek to understand them. All beings within nature should be looked upon as allies.

Work with them and not against them.

Learn and understand the nature of herbs in healing. Learn the types of plants that can provide nourishment. Know how to survive in the wilderness or upon the ocean. Be aware of the plants and animals in any area that you are operating.

Be familiar with the geography and the geology of any area in which you are operating. Is the ground sandy, rocky, or composed of clay? Where are the sources of water? Is there firewood? Which way is the wind blowing? Is there a danger of flooding? Does the air smell of coming rain or snow? Be prepared.

> *"The great thing about a map: it gets you in and out of places in a lot different ways."*
> **- MacGyver: Season One, *The Gauntlet***

When operating in urban areas it is important to know the map. Can you find your way around the city or town? Do you know where the hospital is? The police station? The fire department? Are you familiar with the transportation system? Do you know which areas of town are dangerous to your objectives? Do you have an escape route? Do you have a cover? Do you have an alibi? Is your position adequately protected?

The theory of **SITUATION** means that you must incorporate into your planning the following: distances, danger and security, open ground and narrow passes, high and low ground, and the chances for life and death, injury or capture.

Do you have the means to cover distances? Do you have valid travel documents? Are you prepared for danger? Have you undertaken proper preparations for security? In the event of an accident or attack, have you made arrangements for medical and life insurance? Do you have an up-to-date and valid will? These are very important considerations. Failure to insure yourself will burden your followers and allies.

COMMAND Signifies Leadership

Leaders must possess wisdom, intuition, sincerity, resourcefulness, and benevolence tempered with strictness and courage.

There are nine types of leaders:

There are leaders who guide with virtue and treat all equally and with courtesy. Such a leader is aware of the conditions of followers and takes notice of their troubles. Such leaders are called **humanistic**.

There are leaders who do not avoid any task and are not influenced by profit. Such a leader places honour above self and is called a **dutiful leader**.

There are leaders who do not allow themselves to become arrogant because of their high position. Such a leader does not make much of their victories, but remains wise and humble and continues to be both strong and tolerant. These are **leaders of courtesy**.

There are leaders whose actions are unpredictable and unknowable, whose movements and responses are varied. Such a leader can turn impending defeat into victory. These are the **clever leaders**.

There are leaders whose lives set an example for others. Such leaders are the **saints** of the environmental movement and a source of inspiration.

There are leaders who lead their followers into the field. Such a leader displays courage and skills in confrontation. These are **field leaders**.

There are leaders who face the overwhelming odds and dangers of campaigns, displaying no fear and remaining calm. Such leaders are always in the front during an advance and in the rear guard during withdrawal. These are **frontline leaders**.

There are leaders who have a gift for communication, who understand and can manipulate the media. These are **charismatic leaders**.

There are leaders whose skill, tactics, and courage strike fear and loathing in the hearts of the opposition. Such leaders never spurn good advice and are always both humble and firm. These are the leaders that exhibit the traits of the entire spectrum of leadership virtues. These are uncomplicated men and women yet they have numerous strategies. These are **great leaders**.

TALENTS OF LEADERSHIP

To detect treachery and danger.
To win the allegiance of others.
To be able to rise early in the morning and to retire late.
To speak words that are discreet yet perceptive.
To be direct yet circumspect.
To have courage and the will to fight.
To be of martial bearing and to be fierce of heart.
To understand the hardship of others and to guard against exposing followers to cold and hunger.
To be trustworthy and just.
To be able to understand the mass psychology of the general public and to understand media tactics for communicating with the general public.
To be able to read the signs of the sky, the earth, the sea, and the stars.
To protect all beings and to protect the Earth by understanding that all is one.
To display maternal affection for all living beings.
To lead with the heart, plan with the mind, fight with the body, and triumph with the spirit.

"90 percent of success is *just showing up."*
-Woody Allen

In other words, be active not reactive, get off your ass and get involved. Take the initiative. Stand up and be there.

EVILS OF LEADERSHIP

"Wealth should come like manna from heaven, unearned and uncalled for. Money should be like grace· a gift. It is *not worth sweating and scheming for."*
- Edward Abbey

Never be greedy.
Never be jealous or envious of those who are wiser, more able, or more wealthy.

Never make friends with the treacherous or believe the words of the slanderous.

Never judge others without judging yourself.

Never be hesitant or indecisive.

Never be influenced by flattery or take advice from sycophants.

Never be heavily addicted to alcohol, drugs, gamboling or sex.

Never lie with malice or have a cowardly heart.

Never boast wildly and never talk without courtesy.

Never be close-minded.

SKILLS OF LEADERSHIP

To know the disposition and power of the enemy.

To know how to approach and withdraw from the enemy.

To know how weak or how strong your opposition is.

To know the ways of nature and human psychology.

To know the geography, the terrain, and the situation in the theatre of operations.

THE VICE OF ARROGANCE IN LEADERS

Leaders should never be arrogant. Arrogance leads to discourtesy. Discourtesy will alienate followers and allies. Alienation leads to rebellion.

By **METHOD** AND **DISCIPLINE** you must understand the importance of arranging your forces. You must understand how to delegate responsibility. You must understand maintenance of your forces. You must also understand how to manage your expenditures, your resource base. To properly plan a campaign for victory, the following questions must be answered:

Is your position morally superior to your enemy according to society values and morals?

Which side has the best leadership?

Which side has the advantages of natural and situational circumstances?

Which side is most positively motivated?

Which side rewards its followers better? Which side is most just in administering discipline?

On whose side are the leaders and the followers more highly trained and experienced?

Which side is the stronger?

Which side has the better media strategy?

These considerations will decide in advance if a campaign will win or lose. Within the environmental movement there are four working structures; These are individuals, action groups, organizations, and coalitions.

An individual structure is concerned with the path of the infiltrator and the saboteur. It should be covert and low profile. Leadership is singular and personal, and responsibility involves only the individual.

An Action Group is a small active unit operating on its own or on behalf of an organization. There should be a field leader. If acting on behalf of an organization there should be an organizational leader to whom the field leader reports. Action groups may be low or high profile depending upon the strategy employed.

An Organization must have a good administrator and it must have a clearly defined chain of command as well as a listing of responsibilities. Organizations may only act overtly. Covert action must be delegated to an Action Group.

The most effective political approach is to link organizations into coalitions. The danger here is that the strategy set by a Coalition may only be as strong as the weakness member of the coalition. It is possible however for a stronger group to manipulate weaker member organizations into supporting a stronger approach. Use diplomacy and deception if necessary.

DECEPTION

All confrontation is based on deception. This is called the strategy of tactical paradox.

When you are able to attack, you must seem unable.

When you are active, you should appear inactive.

When you are near, you should have the enemy believe you are far. And when far, near.

Bait the enemy.

Pretend to be disorganized, then strike.

If the enemy is secure, then be prepared.

If the enemy be of superior strength, then evade.

If your opponent has a weakness of temper, then strive to irritate.

Make a pretence of being weak and cultivate your opponent's arrogance.

If your opponent is at ease, then ensure that they are given no rest.

If the forces of your opponent are united, then seek to divide them.

Attack when the enemy is unprepared.

Appear when you are not expected.

The leader who wins makes careful plans.

Even an intuitive strategy without design or concept should be a part of an overall plan.

Preparing plans is an important factor in the Art of Strategy when it is applied to the service of the Biosphere.

Preparation for Death.

> *"The Way of the warrior is resolute acceptance of death."*
> **- Miyamoto Musashi, 1648.**

The Earth warrior, like all warriors, must be prepared for death. Live each day fully as if it were your last.

Do not burden your soul with unfinished business. Be at peace with those you love. Honour your friends and cherish your loves.

Watch your enemies and be vigilant but never surrender to fear.

The knowledge and acceptance of the inevitability of your death coupled with your acceptance of dying will allow you the freedom to proceed and to fight without fear.

> *"Hoka hey"*
> **-Lakota expression meaning: "It's a good day to die."**

Chapter 2

DECEPTION

The Disciplined Art of Strategic Deception

Strategy is a disciplined and complex art.

In this chapter I will focus as an example on my campaigns between 2002 and 2011 into the Southern Ocean to obstruct the illegal activities of the Japanese whaling fleet.

Sun Tzu was the most brilliant strategic thinker in the history of civilization. His book The Art of War has been a guidebook for me for decades and my book **Earthforce!** is a guide to strategy for the environmental warrior and is based on the chapters of Sun Tzu's three thousand year old masterpiece.

My early training in Aikido, Kendo and Western fencing taught me a great deal about strategy and tactics and along with a study of the writings of Miyamoto Musashi, I have over the years developed an approach to campaigns and actions that has allowed the Sea Shepherd Conservation Society to be effective and to survive.

This approach has also incorporated for the modern age, the media strategies and philosophy of Marshall McLuhan.

Modern media is a battlefield of opposing ideologies. Upon this battlefield, truth is manipulated and thus becomes a tool of strategy. There is no objective truth outside of science, and subjective truth is the root cause of conflict. If you control the media, you control a perception of reality. The marketing of perception by the FOX network is different from CNN or MSNBC or National Public Radio. All conflict is merely the result of disagreements in perception of reality. The Fifth Estate is the key to controlling and manipulating perception.

"We both have truths, are mine the same as yours?"
- Pontius Pilate in *Jesus Christ Superstar*

Rules of Engagement

Since the day that I founded the Sea Shepherd Conservation Society in August 1977, we have never yet been successfully sued in a civil court, we have never been convicted of a felony crime and we have never injured anyone nor have my crew ever sustained any serious injuries. Most importantly, we have been effective in defending endangered and protected species and habitats and at the same time raising global awareness of the diminishment of bio-diversity and the destruction of habitat on our magnificent planet.

We have done this by organizing and deploying activist campaigns that operate with discipline and a solid understanding of the media environments.

The foundation of strategy according to Sun Tzu is deception. Sometimes deception is confused with lying but the difference is that lying is unethical whereas deception is strategic. The difference between lying and deception is based on perception. The truth is always a lie in the eyes of the opposition. Strategy requires an adherence to the tenets of disciplined strategy. When weak, appear strong. When strong, appear weak. Never allow the enemy to predict your actions. Through unpredictability and deception comes success in confrontations and campaigns. Focus on the objective and understand that the end justifies the means within the rules of engagement that have been adopted for the campaign. Never be consistent nor inconsistent, never be predictable nor unpredictable, never allow the enemy to understand who you are or where you are coming from. Keep your enemy confused, keep them on the defensive, provoke them, disturb them, upset them, deceive them, confuse them and ultimately overcome them.

Deception against your ally, wife, husband, children, parents or a friend can be quite rightly characterized as lying. Deception against an enemy is simply strategy. Failure to comprehend this is simply a failure to grasp the practical definition of strategy.

Never worry about criticism from your enemies. Criticism from the enemy is an indication of success. When they are in control, they have no need of criticism. Constructive criticism can be accepted from allies and friends with discretion. All other criticism is always defensive and comes from a place of weakness. When your enemy is criticizing you, they are exposing their weakness. The more angry and disrespectful the criticism, the weaker and more insecure is their position. All criticism should be embraced as an indication of success over the opposition.

If your enemies make accusations, ignore them. If your enemy becomes angry, antagonize them. If your enemy becomes conciliatory, don't trust them. The best place to place your enemies is to have them vocal and visible. It is the quiet enemy that one must be vigilant against. Those who talk, rarely act, and those who act, rarely talk.

> *"It is not the critic that counts; not the man who points out how the strong man stumbled or where the doer of deeds could have done them better. The credit belongs to the man who is actually in the arena; whose face is marred by dust and sweat and blood…who, at worst, if he fails, at least fails while daring greatly. Far better it is to dare mighty things, to win glorious triumphs even though checkered by failure, than to rank with those poor spirits who neither enjoy nor suffer much because they live in the gray twilight that know neither victory nor defeat."*
> **- Theodore Roosevelt (1858-1919), 26th U.S. President**

Deception is applied in different ways depending upon the nature of deception. The approach to media is different than the approach to legal or political authorities. The nature of politics is deception. The nature of police work is deception. Media itself is a deception operating under the deceptive illusion of objectivity. Media actually defines reality and therefore it is the ultimate tool of deception. The established media defines reality in a much different manner than the alternative media. The internet is the ultimate application of deception because it allows absolutely anyone to be anyone they like and to advance their own opinions as fact. Facts do not exist, just the perception of what is fact. Reality itself is what one perceives reality to be and everyone lives in a separate reality.

Objectivity is a deception. Sustainability is a deception. Foreign aid is a deception. Scientific research whaling is a deception. Religion and politics are profoundly deceptive. Deception has defined the history of civilization and cannot be denied. However, it can be understood and it can be directed towards the perceived good over the perceived bad. Again this is a matter of perception. However, those who utilize deception for what most people consider evil or wrong-doing seem to prevail more often due to not having any moral qualms about deception.

I experience reality in a manner different than others and others experience reality in a manner different than me and because we all experience reality based on our own perceptions, our own bias and our own life experiences. It is impossible to define a consistent and universal perception of reality. Thus the truth of one person becomes a lie to another. The experience of one person is an abstraction to another.

And thus we cannot expect to be understood for whom we really are, or what we have experienced. All that we can expect is that the objectives that we have set for ourselves with our own lives will be realized through strategy.

The art of strategy is to understand the objective and to not be dissuaded from following the course to that objective and to do so within guidelines set by yourself and your allies and to ignore the rules and the imposed guidelines set by others, especially your enemies.

My objective with regard to Sea Shepherd's annual campaign to the Southern Ocean to end illegal Japanese whaling activities is to sink the Japanese whaling industry economically and to drive them from the Southern Ocean Whale Sanctuary.

I measure this success by whaling profits negated, the lives of whales saved from the harpoons and by an increase in media and public awareness. Towards this end I have no interest in the opinions of pro-whaling factions, nor do I have any interest in the opinions of people who disagree with our tactics. We simply need not concern ourselves with such things. My concerns are to stay within the guidelines of my own self-imposed rules and in this case it is to not cause injury and to not commit a felony crime. Everything else is subjective politics, and politics is nothing more than the opposition of differing opinions.

The implementation of strategy must be focused and not influenced by outside forces including fear of criticism, injury, death or failure. Passion and instinct, experience and discipline, a steadfast determination and a supreme confidence are the characteristics of a successful strategist.

In the pursuit of an objective, the focus must be placed on the primary opposition. For Sea Shepherd in Antarctica that means intervening and stopping the Japanese whaling fleet in the Southern Ocean Whale Sanctuary. Towards that end, we have laid out three guiding rules – our rules of engagement. The first is to not injure any person while in pursuit of this objective. The second is to stay within the boundaries of the laws covering felonious actions. Thus we do not commit capital crimes. Misdemeanor or non-indictable activities are allowed within the context of the actions. Sea Shepherd has had convictions for non-felony actions like breaking the Canadian "Seal Protection Act" that makes it illegal to witness the killing of a seal. The breaking of non-indictable laws is justified on two points. The first point is to challenge the law and the second point is because the particular law is unjust. Justice must always take precedence over laws covering non-capital offences.

The third rule is to not compromise, to not retreat and to not surrender.

Strategy serves the interest of the objective and is not relevant to any other issues or situations not pertaining to the objective if the actions fall within the framework of the rules of engagement.

Whereas the foundation of strategy is deception, the purpose of strategy is to realize an objective. This involves focus, and it involves not only deception, but also discipline, deployment, distraction, diversion, division and documentation. In other words, the 7 "D's" of a successful campaign.

Strategy is like a bow and tactics are the arrows. The preparation and handling of the bow directs the flight of the arrow. The objective is the target.

The Sea Shepherd Conservation Society has undertaken eight campaigns to the Southern Ocean to oppose the illegal activities of the Japanese whaling fleet. With each campaign we became better funded and more experienced. With each campaign we have pushed the envelope a little more aggressively while keeping within the rules of engagement.

We deal with preparations and with supply lines to deploy our ships and crew, we then search for the fleet for the purpose of interception to allow us to obstruct their operations and to prevent the killing of whales. Finally, we document the interventions to deliver a message to the international public. Thus we fight the whalers in the field and in the court of public opinion. Sometimes we need to fight the whalers in courts of law.

To be successful in this effort we must have a solid strategy and a coordinated leadership with a reliable chain of command. But sometimes reliable needs to be seen as unreliable and competency needs to be seen as incompetency and sometimes victory needs to be perceived as failure and failure as victory.

In any campaign, if the supply lines are maintained, if the leadership remains strong and decisive and if the crew remain passionate and determined, victory is a certainty no matter how protracted the conflict. A failure in the supply lines leads to fatigue and demoralization of the leadership and the crew. A failure in judgment and strength by the leadership leads to weakened supply lines and demoralization and fatigue in the crew and the loss of passion and determination in the crew means certain defeat.

Therefore the leadership has three primary responsibilities. First, to address preparations and supply lines, secondly to make decisive decisions regarding preparations, supply lines, recruitment, training and maintenance of the crew and thirdly to decisively remove any weaknesses in leadership and crew when tactically required.

To win the Whale Wars we must have our ships prepared and our supply lines maintained. We have done that and we have gotten better at doing it with each campaign. I think our leadership has gotten stronger with each campaign and I have seen our crews becoming stronger each season by combining experienced veterans with the passionate dedication of new recruits. We need both.

For the critics who say that our crew are not professional, my answer is that I do not want a crew of 100% professionals because professionals cannot be relied upon to tackle the missions that our volunteers embrace with an enthusiastic and passionate approach. We need just enough professionals to keep the ships running and the volunteers alive, but we need the volunteers to achieve the impossible.

Our impossible mission is that we are a small non-governmental international organization sending smaller ships to the Southern Ocean to confront a larger number of bigger and faster ships crewed by whalers quite willing to cause us physical harm. The whalers and their ships are subsidized and supported by one of the most powerful nations on Earth - Japan! We have to locate this fleet in a vast area of one of the most remote and hostile environments on Earth – Antarctica! Once found we must shut down their illegal whaling activities while not causing any injury to any of the whalers while they are trying to injure us.

This is much more complex and difficult than it sometimes appears but the perceived simplicity of the campaign is a deception in itself.

If we keep focused, if we keep our resources strong and wisely deployed, if we maintain a strong leadership and if we maintain passion within the ranks, we will achieve victory over the whale killers by persistence and a passionate commitment.

There can be no retreat and there can be no surrender in any campaign unless it is a tactical retreat.

The integrity of the Southern Ocean Whale Sanctuary must be restored. That is our objective and we will achieve that objective within the boundaries of our rules of engagement meaning we will not injure any of the opposition and we will not engage in a capital crime.

With this strategy we are winning and will win the Whale Wars in the Southern Ocean.

What is, may not be all that is Perceived

Play the fool if thou arte the King,
Play the King if thou arte the Fool.
Weak is Strong and Strong is Weak,
Right can be Wrong, and Strong may be Meek.
What is Right, can be Wrong.
What is Wrong, may be Right.
Our Strategy is what makes us Strong,
The superior Deception wins the Fight.
For in the end, Justice will prevail,
In our Noble Quest to Save the Whale.

Chapter 3

CONFRONTATION

Vi Et Armis*

The longer the confrontation, the greater the expense.

The longer a confrontation lasts, the more followers will be lost from fatigue and disenchantment.

Conflict is like a fire. If you do not put it out, it will burn itself out, but only after it consumes the participants.

If a confrontation lasts too long, your strength will be diminished.

The longer a confrontation lasts, the more dissension there will be in your ranks. Dissension can lead to rebellion by your own forces against you.

Prolonged confrontation is not a wise strategy. Instead, maneuver your opposition into fighting a prolonged campaign.

Keep up a continued token resistance if need be. Expend energy and resources in short bursts only. Attack and retreat, attack and retreat.

Wear down the enemy with hit and run tactics if you find that victory cannot be swift and decisive.

Tree-spiking and monkey-wrenching tactics are excellent hit and run tactics. Strike from the shadows and then disappear into the shadows.

The forces of environmentalism are small. The only strategy for conflict against corporate power, mass media propaganda, and anthropocentric attitudes is a strategy of attack and retreat.

Remember, environmentalists must have no illusion about decisive victory or even victory in our lifetime. We are simply buying time for the future.

Our superior ally is the Earth herself and the forces of apocalypse that she will muster against the oppression of human arrogance.

* Latin: Vi et armis means "By force of arms".

Do not make haste with your plans, but do not delay once your plans have been translated into action.

You must calculate the time and means of your victory in advance.

Being ahead in your timing is the single most important factor in obtaining a victory.

The environmental movement has been, for the most part, a reactive movement. Rarely has there been campaigns where the advance of the enemy against the ecosystem was anticipated and countered before the enemy had moved.

Do not wait for an area to be declared as a target for a forest clear-cut. Spike the trees years in advance. This will counter forest industry and government security measures and it will better disguise the evidence of spiking. In addition, you will be ready to retaliate to an announcement of proposed logging with a media release stating that protective measures for the trees have already been taken.

The struggle to defend the environment is a guerrilla movement by necessity.

Activities by environmentalists are for the most part a series of small, inexpensive skirmishes against a powerful, better equipped opposition. A direct face to face encounter will almost always result in failure because of the superior financial resources of the opposition. A prolonged face to face campaign on open ground will certainly result in failure. The way of the Earth Warrior is to rely on the covert attack, the surprise attack, and planned defence security.

The second most important factor is resourcefulness. You must be prepared to survive in the field. Recruit crewmembers with good foraging and scrounging skills.

My ships are a collection of materials from around the world, all obtained from ports of call and all obtained freely or as inexpensively as possible. From waste bins, from abandoned buildings, from derelict ships, from Naval yards and from supporters. My crew of scroungers, rievers, and resourceful opportunists has saved us hundreds of thousands of dollars in parts, supplies, and essential equipment.

Also, when possible, seize, utilize, and exploit the materials, tools, intelligence, and weapons of the opposition.

Your financial supporters want results, not excuses. They won't support you if you should fail and continue to use the same strategy and tactics. They will not support you for a prolonged period of time unless progress is made. Victories must be achieved. Checkmates win acclaim. Stalemates will bore your supporters. Losses will deplete some or all of your support. In conflict it is important to win and win swiftly.

It is better to win one skirmish after another than to have a prolonged conflict. Hope is built on decision, courage, and action.

It is important that your followers are rewarded for their opposition to the enemy. These rewards may be material or spiritual. If material, the rewards must be distributed fairly.

Never expose yourself to direct interception without first preparing a strategy to escape.

Most importantly, remember that if you are visibly effective and leave yourself defenceless and open to attack, you WILL be taken out by the opposition-without question.

Thus the more effective you become in visibly hurting the opposition, the lower you must keep your public profile. The only other defence is to increase your public profile to the status of celebrity. The opposition must then weigh the advantages and disadvantages of public reaction to your removal. If a major environmental organization is more powerful than the targeted opposition in terms of money, numbers, and political clout, it is important that the group deploy strategies with tact and diplomacy. A crushing victory may result in a moral backlash if an environmental David is perceived as a Goliath.

Be subtle, never heavy handed. Display compassion, sympathy, and understanding. Never come off as uncaring or bureaucratic. Bureaucracy in itself is an enemy. Never allow the infiltration of uncontrollable bureaucracy into your ranks.

Leaders must never surrender control to a bureaucracy. Confrontations must be in the control of a defined leadership. When confrontation begins, the time for democratic discussion of strategy and tactics has ended. All decisions must then be entrusted to a recognized command.

> *"When this ship becomes a democracy, you'll be the first to know."* –**Captain James Tiberius. Kirk**
> (to a crewmember on the *Starship Enterprise*)

Many groups use the method of democratic consensus in decision making. This is not wise, but if this is the way accepted by the group, then it should always be consensus less one. There is a psychology of some individuals who will block a decision simply for the sense of power that it gives them. In fact, some of these individuals often seek out such groups for this purpose. Their priorities lie in dealing with their own psychological problems and not with the issues at hand.

Consensus always arrives at a decision which is most compatible with the weakest individual participating.

Leadership should be organic, springing from the ground with passion, courage, skill, imagination, and ideas. Leadership should give birth to tactics and strategy.

Tactics and strategy define and controlled by a group effort will always be devoid of the passion of leadership and mitigated to the point of ineffectiveness.

> *"You must know something about strategy and tactics and logistics, but also economics and politics and diplomacy and history. You must know everything you can know about military power, and you must also understand the limits of military power."*
> - **John F. Kennedy, June 7, 1961**
> (U.S. Naval Academy Commencement Speech)

In direct confrontation, never compromise. In indirect confrontation use compromise as a tactic of deception only.

The Earth Warrior must never compromise in defence of the Earth.

Do not put your faith in a complete victory. Ours is a struggle that consists of temporary victories. It is our defeats that may be permanent.

Avoid depression and a pessimistic outlook by focusing your efforts on the present. A warrior does best in the "now" and it is the effort expended in the present that determines the future. The past provides lessons, the future provides an objective but the warrior must be firmly rooted in the present to win.

We must fight without thought of winning or losing, we must struggle without thought and without fear of the consequences. We must fight because it is the just and right thing to do, in fact the only thing we can do. If our strategies are sound, our tactics inspired and if we are guided by passion, resolve and courage the future will be determined as best as we can hope for. However, during the struggle hope, fear and glory must not be in our thoughts. The commitment to the present must be total, and the consequences, be they positive or negative must be allowed to grow from the confrontation in response to the effectiveness of the strategies deployed.

It is always strategy that determines the outcome and thus our full commitment must be to formulating and acting upon the strongest and most effective strategy. Nothing else must matter and we must never be deterred or distracted by emotions like fear, doubt, pride, jealousy, hate, love, anger or defeatism.

A warrior must never be angry. A warrior must not hate the opponent. A warrior must never surrender to fear. A warrior should have faith in the strategy in play and be free of doubt. A warrior must be firm and unshakeable as a leader and loyal as a follower. A warrior must be totally committed, unmoved by criticism and accepting of all consequences that may befall without fear or concern.

The objective of strategy is to win and winning is a consequence of effective strategy whereas defeat is a consequence of a failure in strategy. Disregard both possibilities and simply focus on the process – the strategy and nothing more.

The outcome will be determined by the strategy you implement and exercise with total resolve.

Chapter 4

THE ART OF FIGHTING WITHOUT FIGHTING

Vis Consili Expers Mole Ruit Sua *

To fight and conquer in all your battles is not true excellence. Instead, excellence should consist of breaking the enemy's resistance without fighting. In the practical art of war, the best thing of all is to take the enemy's country whole and intact; to shatter and destroy it is counter productive. So, too, it is better to capture an army entire than to destroy it, or to capture a regiment, a detachment, or a company rather than destroying them. In the environmental movement it is imperative that victories be realized without causing or sustaining injuries or fatalities.

Thus the highest form of leadership is to foil the enemies plans; the next best thing is to prevent the unification of the enemy's forces- to interfere with your enemy's alliances, to divide and conquer. The next is to subvert your enemy's strategy. The last resort is to openly engage in conflict.

The worst policy of all is to besiege. To besiege is to prolong the time and to maximize the expenses. This should be avoided at all costs.

The skillful leader subdues the enemy's forces without fighting: they capture without laying siege and they overthrow without lengthy operations in the field. With forces intact, the skillful leader disputes the mastery of the opponent, and thus, without losing any forces, the triumph is complete.

This is the method of attacking by stratagem, of using the sheathed sword.

It is the rule in conflict, that if your forces are ten to one, then surround the enemy. If twice as many, divide into two forces, one to meet the enemy in front, and one to fall upon the rear.

* *Latin: Vis Consili expers moleruit sua means literally. "force without good sense falls by its own weight." A modem translation would be, "discretion is the better part of valor."*

If the enemy replies to the frontal attack, defeat will come from behind; if the attack is from the rear, defeat will come from the front.

If equally matched, offer battle; if slightly inferior in number, then avoid the enemy. If completely unequal in every way, then tactically retreat.

Using guerrilla tactics, unequal numbers may be substituted with stealth, maneuverability, and hit and run tactics to allow you equal or better odds.

The leader is the bulwark of the campaign: if the bulwark is strong at all points, the campaign will be strong; if the bulwark is defective, the campaign will be weakened.

There are three ways that a leader can bring defeat to a campaign:

1. By commanding followers to advance or to retreat, when there are economic or physical barriers which may make it impossible to obey. This is called "hobbling" your followers.

2. A leader of a military, political, or environmental campaign cannot govern their forces in the same way that citizens are governed by a benevolent state. Humanity and justice are the principles on which to govern a benevolent state, but not an army or parts thereof. Opportunism and flexibility are civic rather than military virtues. To govern your forces in a non-military manner will cause restlessness and division within your followers.

3. A leader must not employ officers or delegate authority without discrimination or without heed to military principle of adaptation to circumstances. This shakes the confidence of your followers.

If a leader is ignorant of the principle of adaptability, then they must not be entrusted with a position of authority. The skilful employer of people will employ the wise person, the brave person, the covetous person, and the stupid person. The wise person delights in establishing merit, the brave person likes to show courage in action, the covetous person is quick at seizing advantages, and the stupid person may have no fear of death or failure.

You should not allow your followers to become restless and distrustful. Do not allow anarchy to breed among your followers.

"You can't be faulted for who you hire. But you can be faulted for whom you don't fire."
- **John Paul DeJoria, CEO of Paul Mitchell Inc.**

A good leader must be ready and willing to remove any obstacles within the organization without prejudice.

There are five essentials for victory:

1. You will win if you know when to fight and when not to fight.
2. You will win if you know how to handle both superior and inferior forces.
3. You will win if your followers are animated by the same spirit throughout all their ranks.
4. You will win if you are prepared and your enemy is unprepared.
5. You will win if you have the strategic capacity to not be interfered within the field by your sovereign body, i.e. government, corporate board, etc.

If you know the enemy and know yourself, you will triumph.

If you know yourself but not the enemy, for every victory you win, you will also suffer a defeat. If you know neither the enemy nor yourself, you will be defeated in every battle.

Sometimes it is best to quit the field when your enemy is in full retreat. Allow them to be defeated without adding motivation for them to renew hostilities. Never back an enemy up against the wall and never kick them when they are crippled.

For example, after years of struggle to end the Canadian seal slaughter we achieved our economic objective of a European Union ban on seal products. This effectively ended commercial sealing in Canada. Although seals continued to be killed, the kill figures dropped dramatically although the government was increasing the quota.

The Canadian government was baiting us by raising these quotas because the strategy of the Canadian government was to bring us back to the frontlines in order to give a face for the sealers to focus on. The government needed Sea Shepherd for the sealers to hate. We denied their strategy by withholding our opposition to allow for the economic reality that we achieved to play out. The sealers may not realize yet that they have been defeated but for all intents and purposes the barbaric slaughter of seals for profit on the East coast of Canada will diminish with each new year. We now need only observe and let the fruits of our long campaign see fruition. There is nothing to be gained from flogging a corpse.

Fighting Your Allies

The environmental and conservation movement is composed of numerous individuals, action groups, organizations, and coalitions. These factions promote different agendas and promote different strategies for achieving their objectives.

It is beneficial in any movement to promote harmony so as to direct the greater force against the primary opposition.

If factions have different strategies, they should seek to agree to disagree on tactics and objectives. If this fails, use diplomacy to manipulate a mutual non-aggression policy. If this fails, then use agents to infiltrate, undermine, and if need be, subtlety destroy your opposing "ally".

Unless it is unavoidable, do not publicly battle an ally. This will only serve the interest of the primary opposition.

If your organization or action group is infiltrated by people promoting a different agenda then isolate these voices through ridicule and non-cooperation. Treat them as you would treat the enemy. If the infiltration becomes strong, this means that you failed in your organizational leadership. The solution here is to abandon the organization and regroup.

As an example, in 1990, Earth First! became seriously divided between the original deep ecology faction and those who later joined with a leftist, feminist, and liberal social agenda. The infiltrators went so far as to denounce an essential Earth First! tenet which advocates monkey-wrenching.

Earth First! co-founder, Dave Foreman, turned the tables on the infiltrators by resigning from the organization and founding a new group called Wild Earth. As Foreman put it: "If you live in a mansion, you're going to fight to keep it. But if you live in an

Apache wikkiup, when the vermin move in, well, then you simply move out."

Foreman's move effectively removed the credibility his faction had as the founders of an Earthfirst that was for the Earth first. The usurpers were exposed for what they were and Earthfirst! as a movement was effectively disassembled. What replaced it was another type of organization compromised by an anthropocentric agenda.

Turn all attacks upon you into strategies of retaliation. Use the force of your enemy's charge and turn it back upon your enemy. Use your enemy's own arguments and tactics within your strategy. Your enemy, if understood, will provide the key to your victory.

Chapter 5

TACTICS

Vincam Aut Moriar *

You must always put yourself beyond the possibility of being defeated and then wait for an opportunity to defeat the enemy.

> "Give me a fair ship so that I might go into harm's way."
> - **John Paul Jones**

To secure yourself against defeat lies in your own hands, but the opportunity of defeating the enemy is provided by the enemy themselves. Hence the saying: One may know how to conquer without being able to do it.

Security against defeat implies defensive tactics; the ability to defeat the enemy means taking the offensive. Standing on the defensive indicates insufficient strength; attacking indicates an abundance of strength.

The leader who is skilled in defence hides in the dark; the leader who is skilled in offense charges forth from the sky. Thus, on the one hand, we have the ability to protect ourselves; on the other, to gain a complete victory.

To see victory only when it is within the understanding of the average person is not excellence. Nor is it excellence if you fight and win and you are rewarded. True excellence is to plan secretly, to move surreptitiously, to foil the enemy's intentions, and to thwart their schemes. The victory can be won without shedding a drop of blood.

A clever tactician is one who not only wins, but excels in winning with ease. A clever tactician will find that victories will bring neither a reputation for wisdom nor credit for courage. Victories gained by circumstances that will not come to light will mean that the world at large will know nothing of them.

* Latin: Vincam Aut Moriar means "I will conquer or die."

Your motivation must not be pride. Your entire focus must be on victory. Victory must be the sole objective without distractions, without fear, without hope or desire for reward or recognition.

A master of tactics will be satisfied with the achievement of plans and unconcerned about public reward. This same tactician wins by making no mistakes. Making no mistakes is what establishes the certainty of victory, for it means conquering an enemy that is already defeated.

Thus the skilful tactician will put themselves into a position that makes defeat impossible and does not miss the moment for defeating the enemy.

Thus it is in conflict that the victorious strategist seeks battle only after the victory has been won. The strategist destined for defeat is the one who first fights and afterward looks for victory.

The excellent leader cultivates moral law and strictly adheres to method and discipline. Thus, it is in their power to control success.

APPLICATION OF TACTICS

For the Earth Warrior, tactical approaches will include the following:

    1. Field Tactics
    2. Media Tactics
    3. Psychological Tactics
    4. Educational Tactics
    5. Intuitive Tactics

1. Field Tactics: For environmental activists, these tactics are known as "monkey-wrenching." These are tactics of sabotage, covert activity, and direct action. The recommended manual for the study of field tactics in the environmental movement is *Eco-Defense-The Field Guide to MonkeyWrenching,* by Dave Foreman. These are tactics to be utilized by the Catalyst and the Infiltrator. These tactics require courage, stealth, skill in the practical application of tactics, and the knowledge of laws and tactics of evasion.

2. Media Tactics: In the realm of eco-warfare, the camera is much more powerful than the sword. Documentation and investigation are powerful tactical approaches. A study of the philosophy of Marshall McLuhan is valuable. All eco-warriors should be knowledgeable in the use of cameras, especially video cameras. A reading of Robert Hunter's book *Warriors Of The*

Rainbow will illustrate the advantage of an environmental group skillfully manipulating the media. These are tactics to be utilized by the Communicator, the Catalyst, and the Shaman. These tactics require skills in speaking, writing, photography, media, and control.

3. Psychological Tactics: A foundation in understanding the psychology of humans and human institutions will be valuable in the manipulation and deception of organizations and individuals. It will also serve in implementing the valuable tactic of divide and conquer. These are tactics to be utilized by the Infiltrator, the Communicator, the Shaman, the Artist, and the Catalyst. These tactics require an understanding of human psychology.

4. Educational Tactics: Utilize the media and direct involvement to reach and influence people and advance your position. These are long-term tactics but overall they are the most essential for realizing a permanent change in attitudes. These are tactics to be utilized by the Communicator, the Artist, the Healer, the Shaman, and the Catalyst. These tactics require skills in media, education, and patience.

5. Intuitive Tactics: These are tactics to be found when the Earth Warrior touches the Earth and is given the tactics directly. These are the tactics to be employed when implementing Miyamoto Musashi's strategy of No Design, No Concept.

These tactics are available only if you are intuitively connected to the Earth.

> *"When you realize the value of all life, you dwell less on what is past and concentrate more on the preservation of the future."*
> **- Dian Fossey**
> (Last entry in her journal the night she was murdered)

Intuition is difficult to describe. It can only be experienced. The intuitive warrior is the most fortunate of warriors. Decisions come easier, response is automatic, and there is an absence of stress, fear and doubt.

An intuitive warrior is one who has attained the ability to function independent of the ego, to function naturally and organically with the advantage of fluidity. It is the way of water. Water always flows where it is easiest to flow. Intuition always finds the easiest path of least resistance.

Chapter 6

EARTHFORCE

Vis Consili Expers Mole Ruit Sua *

The control of a large force is the same in principle as the control of a small group; it is simply a question of dividing up their numbers. It is a matter of preparation and organization. Organize a chain of command. In this way, one may command the many. To manage a few is to manage millions. Divide your followers into ordinary and extraordinary. Use your ordinary followers to engage. Use your extraordinary followers to achieve victory.

The environmental movement is composed of moderate and extremist factions. Utilize both. Use the moderates for direct attacks and the extremists for indirect attacks. Use the moderates for overt activities. Use the extremists for covert activities.

Just as nation states maintain both overt and covert control, so must any effective movement operate through the utilization of both covert and overt units.

To ensure that your force will withstand the brunt of the enemy's attack and remain unshaken, use maneuvers that are direct and indirect. Direct methods may be used in joining battle, but indirect methods will be needed in order to secure a victory.

Environmentalists operating as individuals, action groups, or as an organization can be attacked both directly and indirectly by the government. Directly by the police in response to (a) a real crime or (b) a bogus crime or indirectly by the police utilizing a "sting" strategy. The government can also strike indirectly through taxing agencies. Organizations and individuals can come under investigation and harassment by government taxing agencies. Your defence is overtly to keep proper books or covertly to not leave a paper trail. The nature of a government attack can be determined by your own strategy.

If you utilize covert or illegal tactics, you can expect any direct

---

* *Latin: Vis consili expers mole ruit sua means "discretion is the better part of valour. Literally "force without good sense falls by its own weight" - Quotation from Horace's Odes.*

or indirect strategies from government forces, including your own assassination. Your only defence lies in secrecy, careful planning and vigilance.

If you utilize civil disobedience, you will be directly attacked and possibly arrested. Your defence is through established legal channels. You must be prepared.

If you operate openly you need to be prepared for the possibility of being set up. Be prepared to have a defence. Remember, if you operate overtly, keep a record of all activities and maintain accurate accounting. In other words, keep your nose clean. Be prepared to defend yourself in the courts and in the media. For the overt strategy, the courts and the media are the primary battlefields.

If you operate covertly, do not keep any records of real activities. Do not leave a paper trail of any kind. The covert warrior strives to stay off the legal battlefields. The media may be used tactically, but the covert strategy is indirect and difficult to defend if attacked directly.

Direct action speaks louder than words.

Indirect tactics, efficiently applied, are inexhaustible, limited only by the imagination of the tactician.

In conflict, there are two methods of attack: the direct and the indirect. Yet these two in combination give rise to an endless series of tactical maneuvers. The direct and the indirect lead on to each other in turn. It is like moving in a circle, you never come to an end. The possibilities of combination cannot be exhausted.

The good fighting force must be terrifying in its onslaught and prompt in decision.

The good fighting force must go into conflict mean and come out clean.

An effective environmental action group must be ruthless in its pursuit and quick with decisions.

Direction may be compared to the bending of a bow, decision to releasing the arrow.

Amid the turmoil and tragedy of conflict, there may be seeming disorder and yet no real disorder. Amid confusion and chaos, your force may be without head or tail, yet it will still be proof against defeat. Simulated disorder postulates perfect discipline, simulated fear postulates courage, and simulated weakness postulates strength. Hiding order beneath the cloak of disorder is simply a question of subdividing. Concealing courage under a show of fear or timidity presupposes a stock of latent energy. Masking strength with weakness is to be effected by tactical dispositions.

In brief, if you wish to feign cowardice, you must possess considerable courage. If you wish to feign weakness, you must possess considerable strength. If you wish to feign disorder, you must possess considerable order.

If you wish to keep the enemy on the move, you must maintain deceitful appearances, according to which the enemy will act. You must sometimes sacrifice something so that the enemy can snatch at it. You must hold out baits, keep the enemy on the go and then, with a well organized move, lie in wait for a mistake.

During the month of March 1983, I used my ship to blockade the harbour entrance to the Port of St. John's, Newfoundland in Canada. The objective was to prevent the Canadian sealing fleet from departing. Although I had no intention of attacking any sealing ships, I stated that I would ram the first sealer that left the harbour.

It was an aggressive bluff backed up by a solid reputation that I had already made from ramming ships in the past. I had no intention of actually doing it but I was cashing in the credibility of what I had done in the past to give the bluff credibility.

The bluff worked and the sealing ships stayed in port, thus saving 76,000 seals that year.

To keep the government from attacking, I bluffed and deceived and gave the impression of being dangerously unpredictable.

Feigning a move north, I instead moved south and then west, striking at an unanticipated point where I was able to rout three sealing ships and chase them from the seal nurseries.

To do so, I hid under cover of the weather.

It took nearly a week before the Canadian authorities were able to arrest my ship and crew. The arrest however was anticipated and part of my overall strategy to challenge and defeat the laws with which we disagreed, and we succeeded.

The clever strategist looks to the effect of combined energy and does not require too much from single individuals. The leader must however take individual talent into account and utilize each person according to their capabilities. Use people of courage to carry an attack. Use people of caution to defend your position. Use people of intelligence to give counsel. Do not waste talent.

A leader must not demand perfection from the unskilled or the untrained.

A leader must expect perfection from those whose training proclaim them to be masters of their craft.

If an unskilled volunteer makes a mistake, the fault lies with the leader. The leader must oversee or appoint a responsible overseer to monitor the unskilled so as to avoid error.

The responsibility for all error lies with the leader.

A leader cannot demand victory from followers. A skilled strategist prepares properly so as to utilize a situation. It is the situation which must be controlled and exploited.

A good leader will select followers who will best exploit the situation. The leader uses momentum to march followers forward. Moving the force carries the individuals along. Thus the timid move with the brave. The force moves as one.

A leader must depend on opportunity and expediency in seeking a victory.

Take advantage of the situation. There are three situations to look for in this regard: the morale of your followers and of your opposition, the control of the terrain, and how to take advantage of the weakness of your opposition.

When a leader utilizes combined energy, followers become like rolling stones. For it is the nature of a stone to remain motionless on level ground and to move on a slope; if four cornered, to come to a stop, but if rounded, to go rolling downward. Thus the energy developed by trained and well disciplined followers is as the momentum of a round stone rolled down a mountain thousands of meters in height.

Direction of energy and force is the business of strategy. The direction of energy and force is the responsibility of the leader. The direction of energy and force towards the benefit of the Earth is the strategic goal of the Warrior for the Earth.

An Earthforce is an organized fighting force of warriors dedicating their energies towards the objectives of protecting the biosphere. Force and energy plus organization equal an effective Earthforce.

> *"It is not enough to understand the natural world:*
> *The point is to defend and preserve it."*
>
> **- Edward Abbey**

Chapter 7

STEALTH AND UNPREDICTABILITY

Certum Est Quia Impossible Est *

If you desire that your impact be like a tsunami or an earthquake, you must understand the visible and the invisible.

When possible, you must be first in the field. The enemy should come to you. If you are not first in the field, do not rush in to do battle. You must ensure that your fighting strength has not been exhausted. You must impose your will upon the enemy. At the same time you must not allow the enemy to impose its will upon you. Sometimes by holding out an advantage to your enemy, you will draw your opponent to you voluntarily; or, by inflicting damage, you can make it impossible for your enemy to draw near. In the first case, entice with a bait; in the second, strike at some important point that the enemy will be forced to defend.

If the enemy is taking a rest, harass; if stationary, force your opponent to move; if well supplied with supplies, wait and work to cut off the supplies.

Appear at places the enemy must rush to defend, move swiftly to places where you are not expected. Attack where your opponent cannot defend. Hold positions that cannot be attacked. A leader skilled in attack will force the opponent to not know what to defend. A leader skilled in defence will force the enemy to not know what to attack.

Utilize the arts of subtlety and secrecy. Learn to be invisible. Learn to be inaudible. Fight when you wish to fight.

If the enemy is secure, force them to move by attacking at an undefended point. If the enemy wishes to force you to fight, then cut his lines of communication and move around and behind the strong points of your opponent.

If you do not wish to fight, prevent the enemy from engagement.

* Latin: Certum est quia impossible est means "it is certain because it is impossible."

You can do this by throwing something odd and unpredictable in the way.

In 1981, two of my crew and I landed on a Siberian beach to document evidence on illegal Soviet whaling activities. We were approached by two Soviet soldiers. I turned my back on them and stepped into our inflatable boat. My crew sat in the bow facing the beach as I slowly motored away from the shore. "What are they doing?" I asked. My photographer replied, "They have their rifles aimed at us. I think they're going to shoot."

"Smile and wave." I replied without looking back. My two crew smiled and waved and I kept my back to them until we were out of range. They did not shoot.

My tactic confused them. They were uncertain of who we were. We acted friendly. They would be punished severely if they had made a mistake. It is psychologically difficult to shoot a person in the back who has not demonstrated a threat. It is even more difficult to shoot a person who is smiling and waving at you. At the same time, it was a good risk. I decided to gamble with favourable odds and we won.

Discover the dispositions of your opponent while remaining invisible. You can learn to concentrate your forces while dividing the forces of your opponent.

Divide and conquer. Pit your whole against separate parts of your opponent's whole. To strengthen one part of the whole, another part must be weakened; take advantage of this. You must choose the place and the time for confrontation. Always seek to place the burden of reaction on your opposition.

If the enemy is superior in numbers, prevent your opponent from fighting. Scheme and discover the plans of your opposition and the likelihood of the success of those plans. Arouse your enemy and learn the principle of activity or inactivity of your opponent. Force the opposition to reveal themselves. Compare your forces with the forces of your opposition so as to determine your own strength and weakness.

In making tactical decisions, the most important factor is to conceal your plans; conceal your dispositions, and you will be safe from the prying of the most subtle spies.

Produce victory out of the enemy's own tactics.

All people can see the individual tactics needed to conquer, but it is more difficult to see the strategy out of which total victory must evolve.

Tactics are like water; for water in its natural course runs away from high places and moves downward. In confrontation, the way is to avoid what is strong and to strike at what is weak. Water shapes its course according to the nature of the ground over which it flows. The warrior works out a victory in relation to the behaviour of the opposition.

Therefore, just as water retains no constant shape, so in confrontation there are no constant conditions. The five elements- water, fire, wood, metal, earth are not always equally predominant; the four seasons make way for each other in turn. Modify your tactics in relationship to your opponent and you will be a successful captain of tactics and strategy.

"We humans are full of unpredictable emotions"
- William Shatner, Captain James Tiberius Kirk

In 2008, I formulated a strategy to achieve two objectives for Sea Shepherd. First, we needed to retire our ship the ***Farley Mowat*** because of its age and increased operating costs. Secondly, we needed to instigate a dramatic event to put public attention primarily in Europe on the Canadian seal slaughter. This was required to boost support for the European bill to ban seal products.

The plan was called Operation Tar Baby Farley and it was a strategy taken from the old African American tales of Br'er Rabbit.

I sent the ***Farley Mowat*** into the ice of the Gulf of St. Lawrence and taunted the Canadian Minister of Fisheries into attacking us which he did with an over the top RCMP SWAT raid equipped with automatic weapons and military assault gear against an unarmed European crew thus providing dramatic photos for the European media.

I had taken the precaution to command the ship's operation off sight and to put an all European crew of officers on board. This was important to take the attention off of myself as a Canadian and to focus it on my Dutch captain Alex Cornelissen and my Swedish first Officer Peter Hammarstedt.

The Minister fell for the bait and seized the **Farley Mowat** and charged the captain and first officer for violating the absurd "Canadian Seal Protection Act" for approaching too close to a seal slaughter without a permit.

I seized another media opportunity by bailing them out with a sack of Canadian two dollar coins saying that if the government was going to act like pirates we would ransom our crew with "Double loons." (The Canadian dollar coin is called a "loon" making the two dollar coin a double loon.)

The Europeans were outraged and this helped to pass the bill banning Canadian seal products. The captain and first officer were deported and the ship was fined. When I refused to pay the fine, the government responded with a threat to seize the ship. I told them to do so.

In the end the fine was not paid and we successfully disposed of our retired ship with the help of the Canadian government. Both objectives were achieved.

It is important to understand that in this case nothing happened that we did not predict in advance. We anticipated the attack and we let the attack and the seizure work for us.

The Canadian Minister of Fisheries Loyola Hearn attempted to discredit me by saying that I was being cowardly for not being on the ship. He was of course frustrated that I had denied him the opportunity to capture me and that we had laid a trap that he walked blindly into. I successfully taunted him but he failed in his effort to taunt me.

It did help considerably that the Minister was extremely predictable and I was able to use his anger and his volatility as well as his political ambitions against him.

"I've found from past experiences that the tighter your plan, the more likely you are to run into something unpredictable."
— **Richard Dean Anderson, MacGyver:** *The Heist*

Chapter 8

MANEUVERING

Coelum Non Animun Mutant Qui Trans Mare Currunt *

Without harmony in your organization, no strategic expedition can be undertaken; without harmony within your expeditionary force, no conflict can be won.

In a field conflict, the leader receives commands from the organization. Having collected and concentrated all forces, the leader must blend and harmonize the different elements at command.

After this initial organization comes the tactical maneuvering, and this is the most difficult. The difficulty consists in turning the devious into the direct, and misfortune into gain. For instance, to take a long and circuitous route after enticing the enemy out of the way. Or starting out after the enemy to reach the objective in advance of your opponent. This shows knowledge of the art of deviation.

Maneuvering with a disciplined force is an advantage, with an undisciplined force it is most dangerous. Never set a fully equipped force in its entirety to snatch an advantage, the chances are you will be too late. Detach a flying column and sacrifice baggage and stores. Your full force cannot operate without it's provisions and equipment. A flying column can achieve the advantage without baggage and if lost it can be expendable.

Never enter into an alliance without knowing the designs of your potential ally. Do not send your full force into unknown terrain. Always make use of local guides.

During my seal campaigns to Newfoundland in 1976 and 1977, I hired local guides. I had also dispatched spies to Newfoundland months in advance to scout the terrain and the temperament of the local population. Prior to launching the campaign, I travelled to Newfoundland myself to assess the situation. These were invaluable precautions.

  * *Latin: Coelum non animum mutant qui trans mare currunt means "those who cross the sea change the sky, not their spirits."*

In conflict, practice dissimulation. Move only if there is a real advantage to be gained. The decision to concentrate or divide your force must be decided by circumstances. Be rapid. Be compact. Raid swiftly and be immovable.

Be sure that your plans are impenetrable.

Ponder and deliberate before you make your move.

Learn the value of communication through signals.

Learn to maneuver the media and to understand that the media does not simply communicate a message, the media IS the message and the media is also the massage. Individuals are **constantly being manipulated, adjusted, realigned, coaxed, stroked and maneuvered.** It is a two edged sword and can easily be turned against you. Keep the blade swinging away from you.

Your followers' spirit will be at its best in the morning. By noon, the spirit will begin to wane and in the evening it is near gone. Remember this for action. At the same time, impose your will in the evening when the spirit to resist is weakest.

Do not interfere with your enemy if their forces are returning home. A person whose heart is set upon returning home will fight with determination against any attempt to bar the way. Allow them to retreat.

When you surround an opposing force, always leave a way free for your opponent to retreat. Have your enemy believe that there is an escape route to prevent the strength generated by despair and hopelessness. Do not put your opposition in a position to be completely humiliated. Humiliation inspires desperation.

Never press a desperate enemy too hard.

"Battles are won by slaughter and maneuver. The greater the general, the more he contributes in maneuver, the less he demands in slaughter."

-Winston Churchill

Never utilize the same strategy twice in the same situation against the same opposition.

In 1981, I landed an expedition in Soviet Siberia to document evidence of illegal whaling activities. I landed my small crew on the beach, keeping my ship some two miles offshore. My crew began to film and photograph the whaling station in full view of two armed Soviet soldiers. The soldiers did not interfere, they assumed that we must be Soviet scientists or government people. By the time they suspected that we might be illegal, we had obtained the evidence and had returned to the ship. None of my crew were apprehended or hurt.

A couple of years later, the Greenpeace Foundation conducted the same campaign in the same area. Eight people were arrested and charged with espionage. Another crew member suffered a broken leg. Their documentation was confiscated.

What was a workable strategy in 1981 did not work in 1983. The strategy of surprise cannot be used twice in the same place against the same opposition.

"Use anger to throw them into disarray, use humility to make them haughty. Tire them by flight, cause division among them. Attack when they are unprepared, make your move when they do not expect it. Be extremely subtle, even to the point of soundlessness, Thereby you can be the director of the opponent's fate."

- Sun Tzu

Chapter 9

MAINTAINING THE CENTER

Virtutis Fortuna Comes *

Do not camp in difficult terrain. Do not linger in isolated positions. In a difficult situation, use strategy, in a desperate situation you must fight.

There are forces that must not be taken on, positions that must not be contested, and commands that must not be obeyed.

Considerations of advantages and of disadvantages must be blended together.

Reduce the leaders opposing you by inflicting damage upon them; make trouble for them, and keep them constantly engaged; hold out allurements, and make them rush about to any given point.

Entice away the enemy's best warriors and counselors. Recruit traitors and introduce spies to weaken the decisions of your opponent. Foment intrigue and deceit, and sow dissension among the enemy leaders and followers.

Corrupt the morals of your enemy. Disturb and unsettle the mind of your enemy by using strategy.

Resist reliance on the possibility your enemy will not choose to encounter you. Instead be always prepared to receive your opposition. Do not rely on the chance that you will not be attacked. Rely instead that you have made your position unassailable.

There are five faults that may affect a leader. These are dangerous faults and a leader is weakened by anyone of them:

The first is recklessness, which leads to destruction.

The second is cowardice, which leads to capture. The overly cautious leader will also be captured.

The third fault is a delicacy of honour, which is sensitive to shame.

The fourth is a hasty temper, which can be provoked by insults.

* Latin: Virtutis fortuna comes means "good luck is the companion of courage."

The last fault is over solicitude for your followers, which exposes a leader to worry and trouble. Keep in mind that in the long run, your followers will suffer more from the defeat in attaining your common goals than they will suffer in striving to attain these same goals. Keep in mind that your followers will complain. It is in human nature to complain. Satisfy all of the petty complaints and you will take energy away from the primary objective. Followers will not respect a leader who fails, regardless of your kindness. Concern and kindness are forgotten by followers when failure results. The leader answers for failure alone. Victory is shared by all. The leader's responsibility is to the objective first and foremost.

A leader's responsibility to followers is to provide adequate resources for the welfare of followers through delegation of authority. A leader must never become preoccupied with the complaints of followers.

A leader in the field must never give in to demands for democracy.

Decisions made are the responsibility of the leader, the consequences fall on his or her head if negative. If positive, the followers share the victory.

I made the mistake of giving in to my crew's demand to leave the field and went against my judgment to appease them. As a result, my ship and crew were captured by the authorities in a place not of my choosing. The humiliation was mine. The responsibility of defences became mine. Most of the crew refused to participate in raising funds for the defence or doing any work for the defence. They fell back to the position of not taking responsibility. The lesson that I learned then was that decisions can only be made by those who have their head on the block. Responsibility for consequences contributes to correct decision-making in the field.

Chapter 10

FIELD CONFRONTATION

Vi etArmis *

This chapter is for environmental actions that take place in the field. It is meant for those actions which take place in wilderness terrain or urban jungles.

Always exercise forethought and never make light of your opponent.

Camp in high places facing the sun. Make sure that you command the view of your terrain.

When spiking trees for example, always start at the top of the terrain and work down. Always post guards and work out a code of signals.

Do not initiate a confrontation with your back to a river or body of water unless you have a means of crossing the water prepared.

After crossing a river, move quickly away from it. If you must engage your opponent on a river, always approach from upstream. Never move upstream to engage. Never anchor below your enemy.

Never put yourself in a position with your back facing cliffs, swift flowing streams, confined places, tangled thickets, or any other natural hazard.

Always give yourself room to retreat. Within the neighborhood of your camp, survey and routinely inspect surrounding hills, hollows, ponds, thick grass, bushes, and trees. These are locations where the enemy may lie in ambush or spy upon your movements.

When your enemy is close at hand and quiet, they are relying on the natural strength of their position. When the enemy tries to provoke a confrontation, they are anxious for you to advance.

If the encampment of the enemy is easy to approach, beware of a possible trap.

Look for the sudden rising of birds in flight to alert you to the movement of people. Startled wild animals are the best sign of an approaching opposition.

Latin: Vi et armis means "by force of arms."

If birds gather on any spot, it is a sign that the area is free of people. Pay close attention to the signs of nature. Disturbances can easily be detected. The approach of humans can easily be detected. Study and look for these signs.

In urban areas, do not attract attention to yourself by running from an incident. Blend in with the people on the street. Do not appear nervous.

In the wilderness, if the enemy approaches loudly, assume confidence among their ranks. If the enemy is quiet, assume lack of confidence or stealth among their ranks.

To avoid pursuit, do not run.

At the first opportunity, find a means to blend in. That which is obvious is sometimes less suspect.

Once while on board a ship that I was sabotaging, I attracted the attention of the police because of my use of a flashlight. They had seen the flicker from the road. Instead of hiding on board the ship, I immediately went on deck and surprised the two police officers who were just getting out of their car on the dock. I hailed them and greeted them cordially. One officer apologized saying he did not know that the ship had crew on board. He told me that he thought that I was a robber. I thanked him for his concern and told him that I had just arrived and that I was attempting to get a generator working. The police left. I started the generator and completed my work with the benefit of the light and electricity.

Always plan your avenues of escape. Know the terrain. It is also good to have a cover, false identification, and an alibi.

The best rule concerning confrontation is to avoid it unless necessary.

1. Always be fully aware of your environment and always maintain a clear mind.
2. Be invisible within the field of operations.
3. Evaluate the level of risk in every situation.
4. Always be in control of your actions. Abide by your own counsel if you are a leader. Trust in your leader if you are a follower.
5. Never cloud your thoughts with stimulants. Do not work with those who do.

6. If you are provoked, try to avoid a confrontation. Most fights are not worth fighting.

7. Trust in yourself first and others at your own risk and only if they have earned your trust.

8. Always maintain a healthy body. Otherwise physical avoidance may not be an option.

9. Remain focused always on your strategic objective. Practice intelligent avoidance, escape confrontation whenever possible and defend yourself only when you must.

10. Learn and practice the art of deception.

Chapter 11

TERRAIN AND SITUATION

A Frome Paecipitium A Tergo Lupi *

Terrain and situations can be categorized as accessible, entrapping, indecisive, constricted, high, and distant.

Terrain and situations which can be crossed freely and with little difficulty are called accessible.

The advantage lies with the side that can most quickly take up a position on such terrain or seize the first advantage of a situation. The advantage lies with the party that moves most swiftly. On open ground, monkey-wrenchers, survey stake removers, hunt saboteurs, and others must move quickly and in a disciplined manner.

Some terrain and some situations are easy to get out of but difficult to return to. The nature of this position is that the advantage is only yours IF the enemy is unprepared. If your enemy is prepared and you do not win, it will be difficult to withdraw. This type of situation is entanglement and the terrain is entrapping. This is not a good situation.

In 1986, My crew members successfully scuttled half of Iceland's whaling fleet. This tactic worked in Iceland because the Icelanders were not prepared. The tactic would not have worked in Japan where the whalers are prepared for sabotage.

Terrain or situations that are equally disadvantageous to both sides are called indecisive. The enemy will attempt to offer bait to entice you out onto the disadvantaged ground.

Do not go forth. Withdraw and feign weakness.

Draw out the other side into the weak position then turn and strike swiftly and decisively. Do not allow your opposition to choose your targets for you. In British Columbia, the logging companies deliberately announced plans to log pristine and environmentally sensitive areas. Such an announcement was then followed by a

---

* Latin: *A fraote praecipitium a tergo lupi* means literally, "a precipice in front, wolves behind." A more modem meaning would be "between a rock and a hard place."

predictable response. As a result, energy and resources were expended in the direction chosen by the logging companies. Other less pristine and ecologically sensitive areas were ignored by the environmental groups and were logged without opposition. In British Columbia, the logging companies have been implementing a superior strategy. The environmental movement has simply responded to the strategy of the opposition and thus the environmental movement has been channeled and controlled by the logging interests.

On constricted ground like a narrow pass, you can only use this type of terrain if you arrive and survey, fortify, and control the position first. You can then await the opposition. Never pursue or seek out the enemy after your enemy has taken possession of restricted terrain before you. When spiking trees in a mountainous region, you must deploy observers with radios on the roads leading into and out of the area. With the observers in place, tree-spiking can proceed without fear of discovery. If the opposition approaches, all activity ceases and the tree-spikers can remain hidden in the woods. Be prepared to spend time in the forest. Never leave a vehicle in the area.

All activists must be dropped off and retrieved at a pre-determined time. You must control the access points to the area to be effective.

The same rule applies to high ground. Arrive first and you can hold it. Never assault a force occupying a position of height. When camping near a body of water, always stake your tent and supplies on high ground.

With regard to distant ground, your supply lines must be strong and protected. It is important to have capable leaders and followers who can exploit both natural resources and the supplies of the enemy. When followers are too strong and the leaders are too weak, the result is insubordination. When the leaders are too strong and the followers too weak, the result is collapse or chaos. When officers are angry and insubordinate to the leader, the result is ruin. When a leader is weak and has no authority; when orders are not clear and distinct; when there are no fixed duties assigned to officers and followers, the result will be disorganization.

The key to the success of the American revolution was the organizing abilities of General Nathanial Green who understood the necessity of maintaining supply lines. General Washington would not have won without the supply line and attack and retreat strategies of Green.

When a leader is unable to survey the enemy's strength and sends an inferior force to engage a superior force, the result will be destruction of one's followers.

The best historical example of this is the massacre of all of General George Armstrong Custer's men at the Little Big Horn in 1776. Custer failed to survey the enemy's strength.

These are the six guarantees of defeat: neglect to estimate the strength of the enemy, lack of authority, insufficient training, unjustified anger, lack of discipline, and the failure to select trained and disciplined followers.

If the leader knows the terrain and can estimate the opposition, can control the forces, and calculate difficulties, distance, and dangers, there will be triumph. If not, there will be defeat.

If you know the enemy and know yourself, you will not be endangered. If you know the ground, control the situation, and know the weather, you will be victorious.

Decisions of strategy in the field must be made in the field. Decisions in the field cannot be made by bureaucrats who are removed from the situation.

A leader must be unconcerned with fame or material gain. A decision to advance or retreat cannot be made properly unless independent of the desires of the leader. The only desire of the leader that must be cultivated is the desire to triumph over the opposition and to attain objectives.

Leaders must respect their followers.

The environmental movement must respect the terrain of grassroots environmental groups. Large national groups will foment animosity by initiating campaigns without coordination with local groups.

Grassroots groups are the strength of the ecology movement.

Two or more grassroots groups in the same area with similar objectives but with different tactics must agree to disagree on tactical approaches. If they do not then energies will be expended on unproductive dispute over non-objective issues. The opposition will win because the forces have been divided.

If one group does not like another group then it should adopt a policy of mutual non-criticism. If one group does not refrain from criticizing, the group that does refrain will be morally superior and must hold fast to a policy of non-retaliation.

In some situations, it is advantageous for a grassroots group to publicly criticize a large national or international organization on the grounds of bureaucracy, abuse of funds, exploitation of issues, and corruption.

Many large, environmental organizations are reflections of the system. In some cases, the difference in morality, organization, bureaucracy, and self-interest between multinational corporations and multinational environmental organizations are practically indistinguishable.

When an environmental group operates for profit and is insensitive to grassroots organizations, such an ecocorporation should be criticized.

It rarely profits a large environmental group to criticize a smaller grassroots group.

Sometimes the opposition will setup a "grassroots" group in an attempt to co-opt an issue. If this is done, the connections should be investigated and exposed with credibility.

An Earth Warrior must be aware of terrain and situations. The warrior must be able to survey, occupy, secure, or decline to hold a location or situation.

The warrior must be able to control both the situation and all directions of the terrain.

Chapter 12

ATTACKING WITH FIRE

Similia similibus curantur *

Sun Tzu recommends five areas that fire may be employed. The first area is to burn personnel. This is not a method recommended by this book. An Earth Warrior must refrain from deliberately attempting to take life. Fire is unpredictable and uncontrollable and for this reason I do not endorse its use as a tactic. Some may utilize the controlled tactics of fire as described by Sun Tzu. I choose not to do so.

The only fire that I do endorse is the fire of passion and in truth this is the most effective fire of all. It is a positive fire, a constructive fire and a noble fire.

I am including this chapter because I do not wish to censor Sun Tzu and there are real lessons to be learned in his strategies concerning fire.

Of the remaining four areas as described by Sun Tzu: the first is to burn supplies, the second is to burn equipment, the third is to burn weapons, and the fourth is the use of explosives and missiles.

When using fire as a tool, always factor in the strength and direction of the wind, and the dryness of the target area. Many fire attacks depend on the weather.

When a wind blows during the day, it will die down at night.

When using fire to attack, always approach the enemy from downwind.

Always prepare your materials in advance and research your materials so as to apply to your needs.

If you burn enemy supplies in the field, it will damage the opposition. It is most effective to use fire to destroy weapons and equipment.

Fire can be effectively used as a distraction.

There are two types of fire: Controlled and uncontrolled.

---

* *Latin: Similia Similibus curantur literally means, "like things are cured by likes" or "fight fire with fire."*

Fires can be controlled in the form of oxygen and acetylene bottles and this method is ideal for attacks on machinery or stationery metal targets such as billboards or bulldozers. Underwater cutting torches can sink ships.

Gasoline or explosive fires are uncontrollable. Once set into motion, these materials cannot be stopped without extensive damage. The possibility of causing death or injury is a serious consideration. Never utilize explosives without proper instructions in the use of these weapons.

Explosives and fire are unpredictable. Exercise extreme caution. My personal preference is to avoid utilizing both fire and explosives.

The use of fire is meant for situations which demand a radical attack. Fire means maximum damage. The use of fire will provoke anger in your opposition. Be prepared for this anger.

The use of fire is not advisable in overt operations. Fire and explosives are difficult to justify from a public relations point of view. Arson is a serious crime.

Fire and explosives are most effectively used in covert operations.

If fire is used against you, Do not panic. Read the direction of the fire, stay low to the ground, and retreat.

Explosives can be used against you. Always check your vehicle for evidence of tampering before you engage the ignition. If possible, use a remote ignition. Use smoke detectors and fire extinguishers. Be prepared for attack by fire or explosives.

Fire is effective, extreme, unpredictable, and dangerous. Be well versed in utilizing fire for offensive strategies and be prepared to deal with fire from a defensive stance.

Chemicals may be used tactically in some cases. There are chemicals which cause extreme odors that are effective for vacating areas or for crowd dispersal. Pay attention to the direction of the wind if used in the open. Allow for orderly escape if used in enclosed areas.

Smoke may be utilized for the same purpose or for masking an approach. Chemical acids may be used to damage machinery. When attacking chemical plants or facilities, be familiar with the degree of toxicity of the chemicals on the site. Have a protective suit available if there is a danger of a spill.

Our established culture has developed a chemically intensive society. There are over 100,000 different manufactured chemicals in production. The chemical companies are in the front lines of those we oppose. To combat the chemical armada, we must understand and we must employ chemistry. Fire can be fought with fire. Chemicals can be fought with chemicals or the knowledge of chemistry.

If you utilize explosives, it is advisable that you have training, either military or industrial. Lack of knowledge may kill you and others. To kill with explosives intentionally is cowardly and can never be justified within the strategy of a warrior serving the Earth. However, if it is the only defence available, it should be carefully prepared and executed with extreme efficiency.

Fire, explosives, and chemicals should be used tactically only by experts. Fire as a strategy means total destruction of the supplies and equipment of your opposition. In our society, the most vulnerable targets are computers. Strike destructively at the data base of your enemy and you will succeed in inflicting heavy financial and material losses. Power and information are one and the same in a media defined culture.

Constructive Uses of Fire

Always use fire to dispose of papers and materials that may be used as evidence against you. Always scatter the ashes. Learn how to make a fire in the field. It could save your life. Remember, it is a weak strategy that includes poorly prepared tactics.

> *"It seems to be a law inflexible and inexorable that he who will not risk cannot win."*
> — **Captain John Paul Jones**

Chapter 13

THE USE OF INTELLIGENCE AGENTS

Nam Et Ipsa Scientia Potestas Est *

"It is wiser to find out than to suppose."
- **Mark Twain**

The most cost effective method of defeating an opposition is through the intelligent and efficient use of intelligence agents.

Agents cost much less than fielding an expeditionary force in the field.

In confronting your opposition, spare no expense in using agents to investigate the enemy situation.

Success against your opposition depends heavily on foreknowledge of opposition movements and opposition strategies.

The only way of gaining foreknowledge is to infiltrate the opposition with agents. Your agents must be able to discover, to know, and to relay the enemy position to you.

There are five types of agents that can be utilized. These five are native, inside, doubled, expendable, and living agents.

If you can arrange an intelligence structure utilizing all five types of agents, then you will have the ideal arrangement for an effective intelligence at your service.

Native agents are citizens employed from the enemy country.

Inside agents are employees of your opposition who work for you.

Doubled agents are enemy spies that you employ.

Expendable agents are those of your employ who are deliberately given fabricated information.

Living agents are those that are able to return with valuable information.

* *Latin: Nam et ipsa scientia potestas est means "knowledge is power."*

Native or Local Agents

It is wise to recruit citizens of enemy nations to relay information to you. Citizens can sometimes be convinced to do so by appealing to their ideological motives (the most effective), or through financial incentive, or coercion.

Provide your native agents with funds and praise. Provide them with their requests. These are valuable agents. They speak the language and they can move about with more freedom than non-native agents.

Always be prepared for betrayal by a native agent.

I cannot give specific examples of my own use of native agents without betraying the confidentiality of my past and present agents. Suffice to say, in every campaign, I have utilized local people for the following purposes:

1. To prepare reports on terrain, movements of ships and cargoes, locations, relevant individuals, and relevant information.
2. To distribute both information and disinformation.
3. To cause diversions.
4. To provide cover, shelter, and supplies.
5. To translate relevant information.
6. To infiltrate for information and/or sabotage.

Inside or Internal Agents

Inside any organization or government you will find dissatisfied officials. Many of these officials can be enlisted to your cause by appealing to their ideals and desires. Although not completely trustworthy, these agents can provide much valuable information.

A more reliable inside agent is one of your own, one you have trained to infiltrate the enemy ranks.

Reward these people generously.

Once, one of our female agents, dressed seductively, arranged a meeting in Norway with a whaling company official. Presenting one of her false business cards, she introduced herself as a buyer for a major Japanese seafood company. She then negotiated a sale with the whaling company for four tons of whale meat. She carried with her a falsified letter of credit from a Japanese bank and details for shipping. Cash on delivery.

The whale meat was shipped to Panama for trans-shipment to Japan. Our agent then notified another agent who met the shipment in Panama, and he was able to bribe longshoremen to give him access to the containers holding the whale meat.

He disconnected the freezers and jammed the temperature gauges to indicate nothing was wrong.

The meat arrived in Japan in a less than fresh state. The Japanese were not aware that it was arriving. The Norwegians were, to say the least, extremely angry and, most importantly, the operation cost the whaling company a great deal of money.

Double or Converted Agents

When you discover enemy agents working against you, offer to recruit them. Reward them generously. Keep them reporting minor or false information to their other employers. Watch them carefully.

The best use of these agents is to convey false information to the opposition.

On four occasions, I have had government agents infiltrate my crew. On two occasions, I was able to suspect their role and thus I was able to exploit their skills, keep an eye on them, and prevent any damage or leak in plans.

On two other occasions, one agent was caught attempting to sabotage the engine and another agent was able to damage the engine and delay a campaign. On a third occasion, I caught a Canadian government agent stealing files onboard the ship.

Because of these incidents, I suspect all members of the crew and watch them closely until I can determine to my satisfaction that they can be trusted.

It is noteworthy in our case that we are relatively unconcerned with government agents since Sea Shepherd policy is to uphold laws, not to break laws. We are wary however of non-governmental infiltrators.

The Greenpeace Foundation in New Zealand was successfully infiltrated by a French government agent. As a result, the Greenpeace flagship, the *Rainbow Warrior,* was mined and sunk.

Earth First! was successfully infiltrated by agents of the Federal Bureau of Investigation in the United States. As a result, some Earth First! activists were set up, arrested, and jailed.

On the first day of January, 1980, I was forced to scuttle one of my own ships in the Port of Leixoes in Portugal. I had taken a crew into Portugal a few days earlier to steal my ship back from Portuguese officials who had taken custody of it in retribution for scuttling the *Sierra*.

One of my crew was an infiltrator from another environmental organization that was attempting to sabotage us. Upon arriving, we found the ship both looted and disabled. I informed the infiltrator that we would scuttle the ship and head north, each of us separately. After scuttling the ship, my crew and I went to the south. The infiltrator headed north, and because we had left numerous signs of heading north, the infiltrator was apprehended and immediately told the officials that we were all heading north to meet in Spain. This bought us time to reach the south and cross the border into Spain. We escaped.

The infiltrator spent four weeks in a Portuguese jail and then was released.

Expendable Agents

An expendable agent is one of your own people that you send on a mission for one of two purposes:

1. A suicide mission with no or little possibility of escape.
2. An agent who is given false or fabricated information so that they will be deliberately caught, and the information will be made available to the opposition in a credible manner.

Expendable agents must not have knowledge of their real role. They too must be deceived. Expendable agents must not be of much value. They are to be sacrificed as part of a tactical procedure.

Living Agents

These agents are the men and women who form the backbone of your intelligence operation. They should be people of intelligence, but give a foolish or common appearance. They must be strong-willed, active, healthy, physically strong, and courageous. They must be accustomed to dirty work, able to endure hunger, cold, heat, and exposure. They must be able to tolerate humiliation.

Your intelligence agents should be familiars. They should be praised and they should be liberally rewarded.

The most important discipline to insist upon one of your agents is one of secrecy. They must be able to keep their mouths shut.

In choosing an agent, you must look for integrity of character, skills, and experience.

Agents must be managed with benevolence and straight-forwardness.

It is imperative that a leader have a subtle ingenuity in order to ascertain the truthfulness of their agent's reports.

Always be subtle.

Utilize your agents often in all your affairs.

A breach in secrecy must be punished severely.

Always be vigilant for enemy agents.

Converted agents can be used to recruit native and internal agents.

The information acquired from the native and internal agents can be used for setting up an expendable agent.

All this information acquired by the first four types of spies can be used by the living agent.

The purpose of intelligence agents is to acquire knowledge of the opposition.

This knowledge can be best obtained from the converted or double agent. This agent brings information and makes it possible for the others to operate to an advantage. Thus, the converted agent should always be treated with respect.

Intelligence is the most important element of any conflict. Whereas deception is the foundation of strategy, intelligence is the backbone of strategy supporting all strategic maneuvers and tactical operations.

"Knowing a great deal is not the same as being smart; intelligence is not information alone but also judgment, the manner in which information is collected and used"

- Dr. Carl Sagane

THE 36 STRATEGIES

Miyamoto Musashi laid the recklessness of his youth upon the anvil of disciplined study. Before he was able to unsheathe the sword of expertise, he had to first unfurl the scrolls of wisdom and dedicate himself to the study of strategy. One of these scrolls, of ancient Chinese origin, was imperative for focusing the young Musashi on building his foundation for strategy. The author is today unknown. Most certainly the anonymous sage was influenced by Sun Tzu. The study was and is simply known as the "Thirty Six Strategies." I have taken the liberty of translating these strategies into workable formulas for the ecologist activist.

These strategies are:

1. The Art of Invisibility. This is the creation of a situation that, through time, becomes familiar or the creation of a situation that gives an impression of familiarity. Within this situation the strategist may move and maneuver unseen. This means that the activist must blend into the community. The activist must cultivate a routine that trains all observers that there is nothing unusual about his or her movements. This is the art of not attracting attention. It was by this means that the legendary Fox was able to gain access to the head offices of Standard Oil in Chicago. He looked like a painter and his apparent task was to paint the president's office. He proceeded to do so during lunch hour, painting the office with foul-smelling oil and the carcasses of dead birds, fish and frogs.

2. The Helpful Big Brother Scheme. When a weak organization is being challenged by a stronger group, a third organization can come to the defence of the weaker organization, thus gaining trust and loyalty and eventually absorbing the defended group without risk of open hostility. At the same time, the aggressor organization can be weakened and given a bad reputation for aggression. Defender gains benevolent control evolving into complete control. David McTaggart utilized this scheme when he mediated a dispute between Greenpeace Canada and Greenpeace USA in 1979. He emerged as International Chairman of Greenpeace International with absolute authority.

3. The Virtue of Friends. If your organization is weakened, appeal to friends and allies for assistance. If you are weakened as an individual, appeal to your friends. Have your friends and allies bear the brunt of conflict while you gather and replenish your strength. Bear in mind that you have many types of friends and allies. Use your "expendable friends" first and foremost, your valued friends only as a last resort. Reward their contributions generously and cultivate their future possibilities.

4. Be Rested as Your Opponent Wearies. Always force your opposition to expend energy while you maintain your own. Before conflict, you must weary your opponent with diversionary tactics and false alarms that cost you little in resources or energy. Have your opponent come to you. Force your opposition to expend resources without achieving results. Observe and monitor and when your opposition is sufficiently weakened- strike!

5. As Others Drown, Dive for the Pearl. Always use the troubles of others as an opportunity to gain for your cause. As others panic, keep calm and observe the environment for an opportunity to advance. The calm warrior steps adroitly around the burning wreckage, selects what is needed and escapes unseen and undetected.

6. Spreading Manure. Give out false information about your intentions and movements. Leave false clues and signals. Lead your opposition to concentrate defence and offense on fronts more desirable for your advantage. When you say that you will be doing one thing, then do another. If utilized often as a strategy, counter sometimes by doing exactly what you suggest when it is apparent that your opponent believes that you are bluffing.

7. Materializing the Ghost. Create a fantasy before the eyes of your opponent and leave clues and suggestions that will lead your opponent to believe your fantasy is reality. Suggest to your opponent that you have what you do not, that you are stronger than you are. By this means you can create security with little expenditure.

8. The Wizard of Oz. A false front. The illusion of a center of strength. A lightning rod for the concerns of your opposition. Have your opponent concentrate on your center of apparent strength as you attack from the mists and the darkness.

9. The Strategy of Passive Observation. Watch your opponents and if they engage in conflict, observe and do not interfere. Stand ready to walk through the ashes to retrieve what can be salvaged.

10. Cloak and Dagger. Infiltrate your opposition. Gain their trust. Win their confidence. When you have the complete trust of your opponents, you can then move secretly against them. Sam LaBudde used this strategy when he crewed on board a Panamanian tuna seiner for the purpose of documenting the killing of dolphins.

11. The Strategy of Sacrifice. The strategist must be prepared to make individual sacrifices to achieve a greater objective. Leadership of a campaign, a crusade or an organization requires sacrifice on the part of the leader. The objectives must take precedence over all other considerations including personal concerns.

12. The Strategy of Resourcefulness. A strategist must take advantage of any and all opportunities. Utilize available materials and resources and seek to be aware of what is available for your advantage. When operating covertly, always appear calm and behave naturally.

13. Piercing the Veil. When your opponent is hidden and the strategy of your opposition is unknown, the Warrior must initiate an action designed to bait a response. Test your opposition and observe both tactical strength and strategic wisdom.

14. The Strategy of Innovation and Imagination. Do not do what others do. Do not duplicate the strategies of others. Use new tactics and resurrect tactics that have been forgotten. The tactic of tree spiking, which was used by the Industrial Workers of the World in the early 20th century, was later used as an environmental tactic to protect trees in the latter part of the century. You must constantly escalate and change your strategy, especially in a media culture where what is exciting today is old news tomorrow.

For example, ramming a whaling ship was big news when I did it in 1979 but not good enough a few years later. We then had to sink a whaler. A few years after that, we had to sink half a whaling fleet. The media culture especially demands innovation and imagination, and always with an eye for visuals and a flair for the dramatic.

15. The General Wolfe Approach. Back in 1759, General Wolfe defeated General Montcalm at Fortress Quebec by tricking the French commander into leaving the city to engage the British on the Plains of Abraham. Montcalm was defeated whereas he would have been able to defend himself behind the city walls. This is the strategy of inducing your opponent to leave the fortress and come to you. In this way, you select the ground and thus, the rules of the engagement.

16. The Strategy of Patience. Do not press your opponents vigorously. A cornered enemy is extremely dangerous. Pursue with patience. Keep your opposition under observation. Keep your opposition on the run. When their energy has expired-move in for the capture.

17. Trading Glass for Diamonds. Offer something of apparent worth to motivate the production of something of more value. You need nectar to attract bees. You need money to attract money. You need power to attract power. Always project a greater wealth and power than you actually possess.

18. Capture the General to Subdue the Army. When your opposition is overwhelming, concentrate on taking out the leader.

19. Pouring Water on the Powder. If your opponent is powerful and cannot be defeated in a direct confrontation, you must concentrate on destroying resources and hurting the morale of the supporters of your opponent.

20. The Strategy of Confusion. Use confusion to your advantage. Take advantage of confusion among your opposition. Recruit support from amongst the ranks of your opposition by inciting confusion.

21. The Lobster Strategy. When a lobster sheds its shell, it leaves the appearance of being, while the living lobster has moved on. Sometimes a strategist must retain the facade of being in one place after having moved on to another place.

22. Lock the Hen House to Keep the Eggs. In other words, keep infiltrators from your opposition in check. If discovered, pretend that they have been undiscovered. Keep them under observation. If they escape, do not waste energy in pursuit.

23. Keep Your Allies Distant, Your Enemies Near. Never let your allies become familiar with your strength or weakness. Keep your relationship with your allies friendly but formal. Secrecy is your best shield for defence and your best sword for offense. Keep an eye on the movements of your allies. Watch your enemies very carefully and be aware of all their movements. Know your enemy intimately and your allies formally.

24. The Strategy of Manipulation. If you have two enemies and one is the enemy of the other, use the facilities of one enemy to defeat the mutual enemy of you and your gullible ally. After defeating the first enemy, turn the facilities of your "ally" to your advantage in defeating your second enemy on their own territory.

25. Seducing Your Opponent's Followers. Constantly seek to recruit talented followers of your opponent. This strengthens yourself and weakens your enemy.

26. The Art of Indirect Chastisement. If you need to criticize a valued soldier or ally, then criticize an inferior for the same offense and ensure that the valued party bears witness to the scolding.

27. The Strategy of Ignorance. To achieve an objective you may have to feign ignorance, stupidity or incompetence. But you must do so brilliantly. Your performance must not be obvious nor of such a nature that your strategy IS betrayed.

28. The Point of No Return. You should manipulate your opposition into a position of no return by enticement with advantages and opportunities.

29. The Fireworks Strategy. If you can't achieve your aims through facts, then baffle your opposition with bullshit. Deceive with dazzling dramatics, fabricate fantasies fired forth with flair and shower your targets with flamboyance and fiery rhetoric Give the public a circus and contain your message within. Educate through the media of entertainment. Exploit existing myths and create your own myths and legends.

30. The Transformation of Guest Into Host. Sometimes an organization can be taken over by a client. Beware of granting influence to a client, director or interested participating party within your organization. Watch your field agents closely for an indication of a desire for control and influence.

31. The Strategy of Feminine Beauty. This is the strategy that allows women to exploit the weakness in most men to be influenced by beauty and sexuality. Women can be more successful in many situations than a man, simply by using seduction as a strategy. In Africa in 1978, we were able to meet with a government Minister because we had a beautiful woman with our team. He told her anything we wanted to know. This is a strategy that will most certainly be criticized by eco-feminists but in doing so they sacrifice what is probably their most powerful approach to strategy. The feminine is a force that must be taken advantage of and this force includes the seduction of attraction and the appeal of sexual promise. At the same time, male Earth Warriors must be on the alert for this strategy being utilized against them. As an example, the Greenpeace office in New Zealand was successfully infiltrated by a French secret service agent who had some of the males there literally whispering confidential information into her ear. For the female Earth Warrior, all opportunities to gain advantage over male adversaries should be exploited. This strategy can also work for straight or gay men or women seeking to influence a gay man or woman. It is strategically advantageous to always take advantage of any weakness for the benefit of the cause or objective.

32. The Strategy of Vulnerability. When you appear weaker than you are, your opposition may make three mistakes. They may become conceited and complacent, giving you an advantage. They may become arrogant and aggressive, which will enable you to control their destruction. Or they may assume that you are attempting to trap them, which will motivate their retreat.

33. The Strategy of Double Agents. Attempt to compromise agents of your opposition so that they will be forced to work towards your advantage.

34. The Art of Being a Victim. Pretend to be a victim of your own organization so that you might win the confidence and support of your opposition.

35. The Strategy of Orbiting Circles. Never utilize just one strategy. Always use numerous strategies. Change your tactics, innovate new tactics, give your imagination full reign, keep your opposition constantly on the move.

36. The Art of Retreat. When facing superior odds, you must never fight. You have three options. Surrender, compromise or flight. Surrender is complete defeat and is ill advised. Compromise means to be partially defeated and is a last resort. Flight means that you are not defeated. As long as you are not defeated, you have a chance to recover your strength and triumph. Every problem has three solutions. You accept, you ignore or you resolve. To accept is unacceptable. To ignore is unacceptable. The only solution for an Earth Warrior is to resolve the problem. If a solution is not available then retreat from the situation until a solution can be found.

Spaceship Earth

We live on a spaceship!

Like any spaceship it is on a voyage through the vast cosmos on a course that orbits the Milky Way Galaxy. It is travelling extremely fast. Slower than if it were wishfully capable of utilizing the fictional Star Trek starship "warp factor", but fast nonetheless at around 250 kilometers per second or 30,000 kilometers per hours.

Even at this speed, a cosmic year, the time it takes for our planet to orbit the Galaxy is 250 million years. In fact, Spaceship Earth has only made 20 complete revolutions of the Galaxy to date. As passengers, we humans have been on a ridiculously short trip so far. The last time the planet was where we are relative to its last complete evolution was ironically the transition period between the Permian and the Triassic when mass extinctions destroyed millions of species and gave rise to the emergence of the dinosaurs.

Of the five great mass extinction events in the history of the planet, this one was the worse with an estimated 90 to 96 percent of the planet's marine species and 70 percent of land animal and plant species wiped out.

That mass extinction event is known as the "Great Dying". The cause is believed to be the break-up of the supercontinent Pangaea that disrupted the circulation of seawater turning the oceans into a stagnant poisoned swamp of carbon dioxide. It was a period that took the planet over twenty million years to recover from and now we find ourselves on the threshold of an emerging mass extinction event called the Homocene that threatens not just million of species but also our own survival.

Like any spaceship, the Earth has an outer hull and a force field to protect the occupants from radiation from space. It has a powerful engine to drive it through space and to power the life support system. The engine is remotely placed some 93,000 million miles away but that massive nuclear furnace sends power in the form of light and heat to the spaceship within eight minutes and it never fails. We can call it MEGASPEED (**M**agnetic **E**nergy **G**ravitational **A**nd **S**olar **P**owered **E**arth **E**cology **D**rive).

Like any spaceship, it has an internal generator to provide additional heat, magnetic and electrical energy. This power plant is well insulated at the centre of the ship. It has a life support system. This life support system provides us with the oxygen we breathe, the food we eat, the water we drink, and it cleans up our mess. It maintains the climate control system keeping us warm, watered and motivated. It even takes care of our bodies when we die. From birth to death this complex life support system provides everything we need and will ever need.

The life support system for our spaceship is called the biosphere.

The biosphere is a complex system maintained by living organisms working with basic chemical elements in a medium primarily of water.

Water is the essential fluid maintaining every single living cell on the spaceship. It is a fluid in constant movement and constant transition moving from liquid to gas to solid to liquid again in a vast circulatory system that transports nutrients and removes waste in a constant never ending program to sustain all life on this spaceship.

This circulation of water moves from ice to sea to groundwater to the atmosphere and courses through the cells of every living thing. Cold water transiting to warm water in the sea and in the atmosphere maintains the currents, the winds, the rain, the storms, and powers the great gulf streams that pump water through the system like a giant heart reaching every single organism on the planet.

And like any spaceship, there are crew and there are passengers.

We humans are merely passengers. We're along for the ride and we spend most of the time entertaining ourselves. We tend to be unaware of the fact that we are on a spaceship with an essential life support system. We're comfortable. In fact, we're so comfortable and so powerful in numbers that we spend a great deal of time simply increasing those numbers to the point where it is beginning to cause irreparable damage to the life support system of Spaceship Earth.

More accurately, we are killing off the crew that maintain this life support system. In fact, we don't really pay much attention to the crew at all. We treat them like they are beneath us, sometimes with disgust and most times with apathy.

Yet without them, we could not exist.

The bacteria, the seaweeds, the plankton, the trees, the flowering plants, the worms, the bees, the ants and beetles, the flies and the fish.

These are the essential beings that maintain the life support system. These are the species that make up the crew of spaceship Earth. These are the species that nurture and sustain us.

And we are killing them, destroying their carry capacity, eradicating them, driving them to extinction without any thought, without remorse or guilt, without even much recognition of what we are doing.

We view them as insignificant, as unimportant, despite the fact that these "lowly" creatures that we take for granted in so many ways make it possible for us to be alive and healthy on Spaceship Earth.

In reality they are more important than us.

Worms are more important that human beings. Bees and ants, trees and fish are more important than we are.

Why?

Because we need them to survive. They do not need us.

Mammals and especially humans contribute very little to the actual maintenance of Spaceship Earth. In many ways, humans are sentient weeds. We may look pretty, we may breed like yeast, and we may believe we are the greatest creation under the stars, but the truth is that we exist by virtue of the fact that billions of creatures from millions of species make it possible for us to exist at all. Without them there would be no art, no music, no poetry, no science, no civilization.

A world without worms would be a world without humans. A world without just a few species of bacteria would be a world without humans.

We even fool ourselves when we refer to ourselves as individuals. We are nothing of the kind. Every human being is a symbiant, or a community of species.

Inside and outside of our bodies dwell a trillion living bacterium, in total making up a kilo of our body weight. There are between 800 and a 1,000 species of these microscopic creatures making themselves at home in and on us.

They are not parasites because we need them and they need us. They digest our food, they manufacture vitamins for us, they protect us from destructive bacteria and they even groom our skin and our eye lashes.

They are as much a part of us as our eyes and skin.

There is a saying that no man is an island entire of himself. This is true in more ways than we think for every person is a nation containing a trillion individual life forms.

We are intimately connected to Spaceship Earth. If we touch or pull on any part of this planet we find everything is tied to everything else.

We are Earthlings living on a water planet and the term Earthling encompasses every single species that dwells on this magnificent blue and white sphere, this beautiful ship of life that sparkles like a jewel in an otherwise barren solar system speeding around a vast galaxy of a hundred billion stars, millions of which also power spaceships similar to us with life forms unique to the circumstances that life evolved on each and every one.

All of these spaceships are grand natural experiments. Many will self destruct and many will evolve further as they journey through the immenseness of the cosmos.

The question for us, is how long will humanity survive on a spaceship as ruthless passengers intent upon the mass murder of the crew of spaceship Earth?

AFTERWORDS

In presenting this book on strategy, I have attempted to construct a foundation for the individual warrior.

The wisdom of Sun Tzu is meant to guide you. The strategy that needs to be developed for your individual needs and objectives must originate from within your intuitive soul and from no one else. Your intuitive strategy must be tempered by the fires of your own imagination and driven by the strength and power of your will.

Remember, always, that the strength of your will, the flames of your imagination, and the spirit of your intuition cannot triumph over the physical laws of the universe.

For every action there is an equal and opposite reaction. Thus the karmic laws must be respected.

Once the law of karma is understood, it becomes safe to move onto the next step of obtaining freedom.

Shadows can only pretend to be warriors. Shadows may go through the motions of a warrior, but they only mimic the true warrior. Humans who are enslaved to security, materialism, emotions, pleasure, pain, abuse, arrogance, appetite, and to dependence are only shadows.

We are locked into our suffering, and our pleasures are the seal. Only by utilizing the charts of imagination can we navigate our discipline to overcome the shadowy temptations of anthropocentric reality.

The first step to freedom is to transcend the anthropocentric reality. Place yourself spiritually beyond the reach of the pseudo-important myths of the self-centered human cultures. Once you have become spiritually armored, you can be free to mindfully participate with the folly of humanity.

A warrior is a being of power. Power can only be realized by discipline. Discipline can only be maintained by the will. The will must be fueled by the imagination. The imagination derives its power from the intuitive soul.

For your teachers, look to both the human and nonhuman inhabitants of the Earth for your lessons. Learn by intuition to see wisdom and reject ignorance.

Above all, do not mire yourself in the muck of pettiness. All individual problems are resolvable, either by solution or by acceptance.

Do not concern yourself with what others might think of you. The strategy of an Earth warrior activist is to say unpopular things to piss people off, to undertake actions which anger and disturb people, and to boldly strike at issues and ideologies where none have dared to strike before.

> "The only thing worse than being talked about, is not being talked about."
>
> - Oscar Wilde

It is the duty of the Earth Warrior to use actions, voice, pen, and camera to provoke thoughts, and to motivate change. The warrior is a catalyst for change. The actions, objectives, and the exemplary life style of a warrior are more important then the mortal ego of the individual.

In choosing a warrior's way and dedicating my life to protecting and serving my planet, I have been richly rewarded.

Now, at the age of sixty, I can honestly say that I have been very much satisfied with my life and achievements. I have saved lives. Because I acted, thousands of wolves and bison, thousands of whales and redwoods, tens of thousands of dolphins, and hundreds of thousands of seals have survived and have been given an opportunity to perpetuate more of their own kind.

If wolves, whales, dolphins, and redwoods survive the hominid holocaust, it will be in some small part because my crew and I had the audacity to intervene in the selfish affairs of anthropocentric humans.

> "The man who is right is a majority. He who has God [Earth] and conscience on his side, has a majority. Though he does not represent the present state, he represents the future state. If he does not represent what we are, he represents what we ought to be."
>
> - Frederick Douglass

Remember, always, that it is the nature of a warrior to act. Do not be daunted by the formidable strength of the opposition. Do not be depressed by doom and gloom predictions. A true warrior must welcome challenge and transform the impossible into the possible.

Because you live in these trying times, it is your task to confront situations created by human ignorance and apathy, and focus your actions through love for the future and all the children of all the children of all species.

Do not be embarrassed to howl with the wolves. Do not be afraid to swim with the sharks. Do not be so proud as to ignore the art of the spider, the wisdom of the whale, the strength of the ant, or the beauty of the planet.

Above all, re-connect with the Earth. Feel the mud between your toes, smell the fragrance of the forests, dance with the stars, and exalt in your animalism.

Eat at the table of the Earthly mother. Drink when you are thirsty and eat when you are hungry. Give nutrients to the forest floor when you are able to do so.

Sleep in the arms of the Earth with the stars above your head when you can. Take every opportunity to do so.

Laugh when you are happy. Cry when you are sad. Scream forth your joy at the climax of sexual unification.

Be an animal, a primate, the outrageous monkey cousin that you are. Be alive and live, live, live.

The love of life inspires passion for life which motivates the passionate defence of life. And in a most ironic way, such passion for life allows the peaceful acceptance and understanding of the value your death has in the vibrant dance of the Continuum.

Environmental activists may be a nuisance and a pain in the ass to the established authorities of the present. However, to the establishment of the future, we will be honoured ancestors.